CHESAPEAKE
PERSPECTIVES

DECODING THE
DEEP SEDIMENTS

The Ecological History
of Chesapeake Bay

D1823670

GRACE BRUSH
Johns Hopkins University

Sea Grant

A Maryland Sea Grant Publication

The ideas and opinions expressed in this monograph are entirely those of the author and do not necessarily represent the views of the Maryland Sea Grant College, the University of Maryland, or the National Oceanic and Atmospheric Administration.

Publication Number
UM-SG-CP-2017-01

The publication of *Chesapeake Perspectives* is made possible in part by a grant to Maryland Sea Grant from the National Oceanic and Atmospheric Administration, Department of Commerce, through the National Sea Grant College Program. Grant number NA14OAR4170090

Cover photograph by Skip Brown.
Book and cover design by Sandy Rodgers.

For more information on this and other publications, or about our program, contact:

Maryland Sea Grant College
University of Maryland
4321 Hartwick Road, Suite 300
College Park, Maryland 20740
www.mdsg.umd.edu

Library of Congress Cataloging-in-Publication Data

Names: Brush, Grace Somers.
Title: Decoding the deep sediments : the ecological history of Chesapeake Bay
  / Grace Brush, Johns Hopkins University.
Description: College Park, Maryland : Maryland Sea Grant College, University
  of Maryland, [2017] | Series: Chesapeake perspectives | "A Maryland Sea
  Grant Publication." | Includes bibliographical references.
Identifiers: LCCN 2016051335 | ISBN 9780943676760 (pbk.)
Subjects: LCSH: Water quality--Chesapeake Bay (Md. and Va.) | Coastal
  sediments--Chesapeake Bay (Md. and Va.) | Ecology--Chesapeake Bay Region
  (Md. and Va.) | Paleoecology--Chesapeake Bay Region (Md. and Va.) |
  Chesapeake Bay (Md. and Va.)
Classification: LCC TD223.1 .B78 2017 | DDC 551.3/540916347--dc23
LC record available at https://lccn.loc.gov/2016051335

Contents

Foreword

Beginning in the late 1970s, research on the ecological processes regulating the Chesapeake Bay intensified with a five-year, $27-million study authorized by the U.S. Congress. Declining water quality and the loss of aquatic life spurred the research to identify the causes — its aim: to begin developing science-based resource management policies that would help reverse the Bay's decline so that this complex ecosystem could recover a level of health for sustaining fish and shellfish populations.

Many studies, then and now, focused on causes for the widespread loss of underwater grasses (submerged aquatic vegetation); the extensive depletion of dissolved oxygen in bottom waters; the apparent spread of harmful algal blooms; the role of pollutants coming from sewage, factories, and farms; diseases that prevented Bay oysters from sustainable replenishment; and the depletion of fish species, such as sturgeon.

While land use was suspected as contributing to the continuing decline of water quality — transforming forests into agriculture and urban lands with consequent runoff of excessive sediments, nutrients, and chemical pollutants — just how much of a factor was it? In what ways had the function of the Chesapeake landscape changed since settlement began? In order to compare the pre-Colonial Bay — close to its "natural" functioning — with the post-settlement Bay, we needed a rigorous understanding of what the Chesapeake Bay was like under pre-settlement conditions. No written records existed, only anecdotal ones. Fortunately, an outstanding scientist, Grace Brush, interested in pursuing these types of questions, was living in Baltimore.

Bay sediments, she argued, harbor records of what the pre-Colonial Bay was like, if we could "read" them. And for more than thirty years, Brush and her many students at Johns Hopkins University have been doing just that — doctoral and graduate students, as well as many undergraduates. Now in her ninth decade, she is still going strong, teaching undergraduate classes and maintaining an active research lab.

In this monograph she explains the nature of her research, the discoveries she and her students have made, and how their findings have contributed significantly to our understanding of how land use affects the Chesapeake Bay system, and in particular the shifting over time from a Bay dominated by benthic processes to one now dominated by pelagic processes.

It is especially gratifying for the Maryland Sea Grant program, and me personally, to publish this monograph. From my own background as a modern processes sedimentologist and a woman scientist, Grace Brush has long been a hero of mine. She and her students have consistently led the field of paleoecology. Here, Brush presents a summary of her Chesapeake Bay research, what we have learned to date, and how her students have been so instrumental in this ground-breaking work. But, additionally, she discusses her path to achievement as a woman scientist.

Brush came to aquatic research at a period when it was nearly impossible for women scientists to get positions in marine and estuarine laboratories — but she was driven to explore and, in retrospect, she was a pioneer. Today, many take for granted that most women scientists are on an equal footing with men. I asked Dr. Brush to write about her scientific life, how she came to do the paleoecological research itself but also the paths she had to travel, while married to a scientist, raising children, and pursuing funding. I think all readers will admire her remarkable story, her fortitude, and, not least of all, the students she has attracted to continue this work.

Fredrika Moser
Director, Maryland Sea Grant

Dedicated to the memory of
Lucien Brush (1929-1994)

Introduction

Over the last three centuries, human alterations of land in the watershed drained by the Chesapeake estuary have led to profound changes in water quality and aquatic resources. What are the changes, how did they happen, and are they reversible? There are no clear answers.

Rich biological resources originate in complex ecosystems, where cause and effect are obscured by a vast interplay among a myriad of organisms. Each of the Bay's organisms has its own lifestyle. Some crawl along the bottom of the estuary; others swim in the water. Some live in brackish regions, some in fresh water, and others in both during different life stages. Generation times range from a day to a century. Organisms feed on each other in a complex food web; the entire system is affected when any component of the living network encounters difficulties. Cause-and-effect questions are still more difficult to address because the water and the land it drains are so intimately connected. Fresh water enters the estuary through thousands of tiny tributaries and many large rivers, in flows intermittently large and small. Salt water is driven into the estuary by the daily tidal cycle. Both fresh and salt water are conduits for materials from the land and the ocean. In addition, the dynamics of this system are complicated by its long history, extending over thousands of years.

To understand how we have altered the Bay within our lifetime and to what extent restoration strategies may be achievable, we need to know its history. As humans, we do not relate to a history of several thousand years, because our time span is extremely short compared with many earth-related changes; we imagine that since the Chesapeake Bay has provided oysters and shad and rockfish for a century or more, it will (or should) continue to do so forever. If the harvest of a species is reduced by disease or overfishing, we set out to "fix" the problem, often not realizing that the change occurred at a particular time in the natural evolution of the system and involves a host of other organisms. If the changes occurred at another time under different conditions, maybe drier or wetter or colder or warmer or stormier, the outcome might have been entirely different. The response of estuarine organisms to human activity is within the context of their response to the ever-changing land and water on time scales much longer than those caused by human forces.

While no written records of the pre-Colonial Bay and only scattered records of post-Colonial history are available, unwritten records are preserved in the sediments where organisms and materials from within the Bay and the surrounding land and air are buried. This paleoecological record of several thousand years allows us to compare changes made by humans with geomorphological and climate changes before humans became a controlling force.

The Chesapeake Bay and Its Watershed

The Chesapeake Bay is the largest estuary in North America and one of the largest in the world. It is situated on the east coast of the USA between 39045' and 36050' North latitude. The Bay is approximately 300 km long with at least 25 major tributaries and many smaller ones draining a total area of about 166,000 km².

The watershed drained by the Chesapeake includes Maryland, Virginia, West Virginia, Pennsylvania, New York, and the District of Columbia. Lands in the watershed are diverse, embodying parts of the Coastal Plain, Piedmont, Valley and Ridge, and Appalachian Plateau physiographic provinces. The Coastal Plain is topographically flat and underlain mainly by unconsolidated sedimentary deposits. The Piedmont, characterized primarily by thick saprolitic soils weathered from igneous and metamorphic rocks, has somewhat higher elevations. To the west, the mountainous Valley and Ridge and Appalachian Plateau provinces formed on limestones, sandstones, shales, and siltstones. Rainfall approximates 100 cm per year, with the greatest precipitation in the spring, though annual variability can be high. Mean annual temperatures range from 11°C to 15°C in the Coastal Plain and Piedmont and from 9°C to 11°C in the Appalachian province.

Geological History

For most of the last two million years, glaciers covered at least half of the Earth, their advance and retreat guided by astronomical forces that control the relationship between the Sun and the Earth and hence the distribution of heat to the Earth. The time period known as the Ice Age consisted of 100,000 years of full glaciation separated by short interglacial periods of

10,000 to 20,000 years. As the earth warmed, glaciers retreated, sea levels rose, and flooded coastal areas gave shape to new landforms, including estuaries.

Seismic profiles of the bottom of the Chesapeake Bay by the Maryland Geological Survey provide partial evidence for three estuaries prior to the current Chesapeake. These data show the earlier estuaries flowing east of the present Chesapeake, discharging into the Atlantic Ocean north and west of Norfolk (Colman et al. 1990) (Figure 1).

Today's Chesapeake Bay began when the most recent continental glaciers began retreating from south to north about 18,000 years ago. Until then, the Susquehanna River discharged directly into the Atlantic Ocean in the vicinity of Norfolk, Virginia. As the glaciers retreated, becoming increasingly smaller, ocean waters continued to rise, flooding the Susquehanna River and its tributaries with saline water. Salinities range widely in

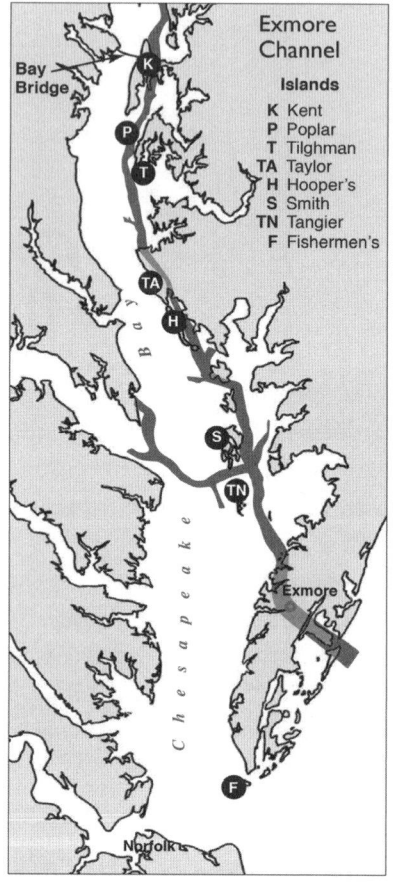

300,000-500,000 years ago

Exmore Channel

Bay Bridge

Islands
K Kent
P Poplar
T Tilghman
TA Taylor
H Hooper's
S Smith
TN Tangier
F Fishermen's

these regions depending on precipitation, and can vary annually in any one area. In dry years, when rainfall is low, less fresh water is discharged from the rivers into the estuary than in wet years, and salt water is moved farther upstream by the twice-daily tides. The opposite happens in wet years, when fresh water flows farther downstream. Hence, salinities range from 0 parts per thousand (ppt) salt at the northern extremity, to 10 to 20 ppt in the Bay's mid-region, and 35 ppt at the southern terminus where the estuary enters the Atlantic. Over the last 10,000 years resilient organisms have adapted to new habitats with a wide range of changing salinities and warming temperatures.

As the glaciers advanced, plants on the land migrated in front of the ice

150,000 years ago   **18,000 years ago to the Present**

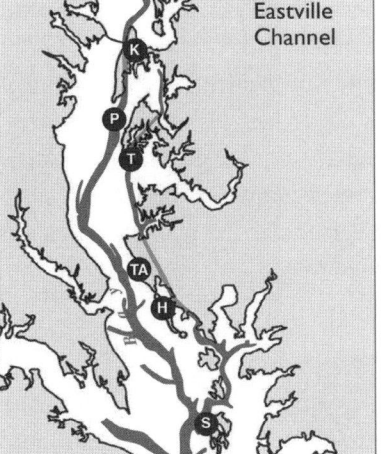
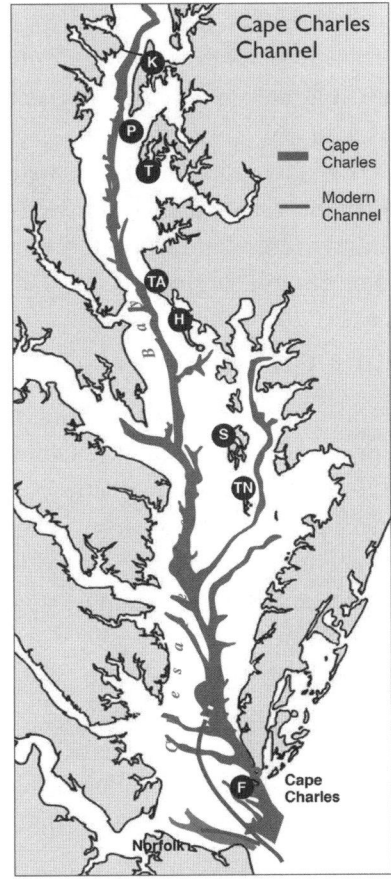

Figure 1 (opposite page and above). Three ancient channels lie buried beneath the modern-day Chesapeake Bay, each one named for the town near where it once exited to the ocean. The Exmore Channel (opposite page) is the oldest buried paleochannel of the Susquehanna River. When the channel was an active river, it was fed by the Patuxent and the Potomac Rivers. The Eastville Channel (above left) has a track that is shifted to the west for most of its run. The Cape Charles Channel (above right) is shifted still farther west. It was the axial channel of the Susquehanna River when rising sea levels began turning the river into our current Chesapeake Bay. Its track runs near the modern deep channel. As a river, it captured all the western shore rivers except the James. Once 160 feet deep at the mouth of the Chesapeake, the channel was quickly reburied under Fishermen's Island by sand and sediment from the sea. Illustrations from Chesapeake Quarterly *magazine, vol. 10, No. 1, redrawn from Colman et al. (1990).*

to areas not reached by the glaciers. Some plants became extinct where migration routes were absent. As glaciers retreated, some species, such as beech, which migrated to the Continental Shelf (dry land during full glacial time), migrated westward. Species of pine and oak that lived out the glacial period in the Gulf of Mexico region and farther south migrated in a northerly direction behind the melting glaciers. The species that occupy the land in the Chesapeake watershed are assemblages of plants that migrated at different rates and from different directions as the glaciers retreated and climate became warmer.

Human History

The human history of the Chesapeake can be divided roughly into two major intervals: pre-Colonial settlement and post-Colonial settlement. Several thousand years before European colonists began arriving, the Bay was influenced primarily by geomorphologic and climatic processes. There is no evidence that Native Americans, whose numbers were relatively small in this area, caused major changes to the surrounding land, such as clearing of forests, nor to the Chesapeake ecosystem: in general, they used the resources when and where available in ways that we might call "adaptive management." The post-Colonial period includes the last 300 to 400 years, during which time population growth and massive changes to the landscape became the dominant influence on both land and water. Humans had forged a new relationship with the environment.

Historical records indicate that when Europeans arrived in the late 17th century, the land drained by the Chesapeake was primarily forested except for a few serpentine areas supporting grasses and other herbaceous species together with freshwater marshes (many created by beavers), brackish marshes along the coasts, and Indian villages. Historical documents also record how rapidly the land was changed as colonists, whose main activity was agriculture, increased in number. Forests were cleared for cropland, buildings, firewood, roads, and other needs. By the early 18th century, 20% of the forests were cleared; by the early 19th century, 60% had been cut; and by the middle to end of the 19th century, over 80% of all forest cover had been removed and replaced by agricultural fields, primarily tobacco and wheat. Later, the increasing imperviousness of land surfaces due to roads and urban structures led to still more changes (Figure 2). This transformation of the land from forest to farms and hard surfaces and the rapid rate at which it occurred was one of the great ecological phenomena in human history — it would have enormous repercussions for the estuary that drained these altered lands.

Forests in the Chesapeake Watershed

Forested land is an intricate system, the patterns of which are governed importantly by a geologic/soils template established millions of years ago — long before the forests evolved and the estuary formed. In the early 1970s, before I began studying the ecological history of Chesapeake Bay, the state of Maryland's Power Plant Siting Program (PPSP) expressed the need for a map of Maryland's forests: knowledge of the composition and distribution of forests throughout the state would be useful for the siting of electric utilities and other needs. The state Forest Service had inventoried tree cover at different times, but these lists were restricted mostly to species important for the lumber industry and did not capture the ecological diversity of the land.

While Forrest Shreve's *The Forests of Maryland* (Shreve et al. 1910) recorded all species on the different soil types in the late 1800s after most of the forests had been removed, it was not a map in that the boundaries of plant associations were not drawn. I proposed mapping the natural forests of Maryland from randomly selected field samples and then overlaying the species distributions with the state's geologic, topographic, and soils maps, hypothesizing that tree distributions would be closely related to the geology and soils from which they obtain water and nutrients.

The logistics of producing the map in three years were challenging to say the least. Making a map that records plant species involves fieldwork. Since Maryland includes parts of four physiographic provinces, we organized the fieldwork into three four-month summer seasons. Because this work was done before the advent of computerized GIS, defining the boundaries of the forest associations for the map, which were based on species with discontinuous distributions, required working with multiple mylar overlays.

Field studies were completed in 1978, and the map was drawn at a 1:250,000 scale, similar to the Maryland state geologic and topographic maps. The map showed a mosaic of 18 different forest types identified by "indicator species" closely aligned with the hydrologic characteristics (wet/dry) of geologic and soils substrates. The state underwrote a print run of 8,000 copies for distribution by *Ecological Monographs* (Brush et al. 1980) and for use statewide. Conservation and other groups throughout Maryland have used it widely.

The study provided important ecological information for a large part of the Chesapeake drainage area, and for interpreting the paleoecological studies of the estuary, which came later. Afforestation was occurring wherever land was not being farmed. A comparison of my studies with those by Shreve showed a remarkable similarity. Since the late 1800s (the time of Shreve's study), when the land was 80% deforested to the 1970s (the time of my study) when deforestation was about 60%, the species reoccupying the land were the same in similar locations, except for a couple of species lost to extinction or undergoing extinction (namely the American chestnut in the 1930s and more recently the American elm). Both studies showed clearly the resilience of the forest-substrate mosaic and provided insights with respect to land-water interactions.

Precipitation on a forested landscape either evaporates or soaks into the ground. When trees are cut, land is plowed, and roads are built, most of the rain runs off the land carrying with it sediment particles loosened from the soil which are deposited in the estuary. Much of this sediment is silt and clay, which has attached to it nutrients, trace metals, and substances derived from weathering rock and decaying vegetation. Also attached to the silt and

Figure 2 (opposite page). Average deforestation rates in the watershed for four periods: pre-Colonial, early Colonial, intensive agriculture, and the modern era (redrawn from Brush 2009). The increase in forest cover from the mid-20th century to the present is due, in part, to the decline of farmland; while much of that land has been developed, undeveloped land has been reverting to forests, at the same time that state conservation efforts have aimed at improved forest management.

Average Deforestation Rates, Chesapeake Watershed

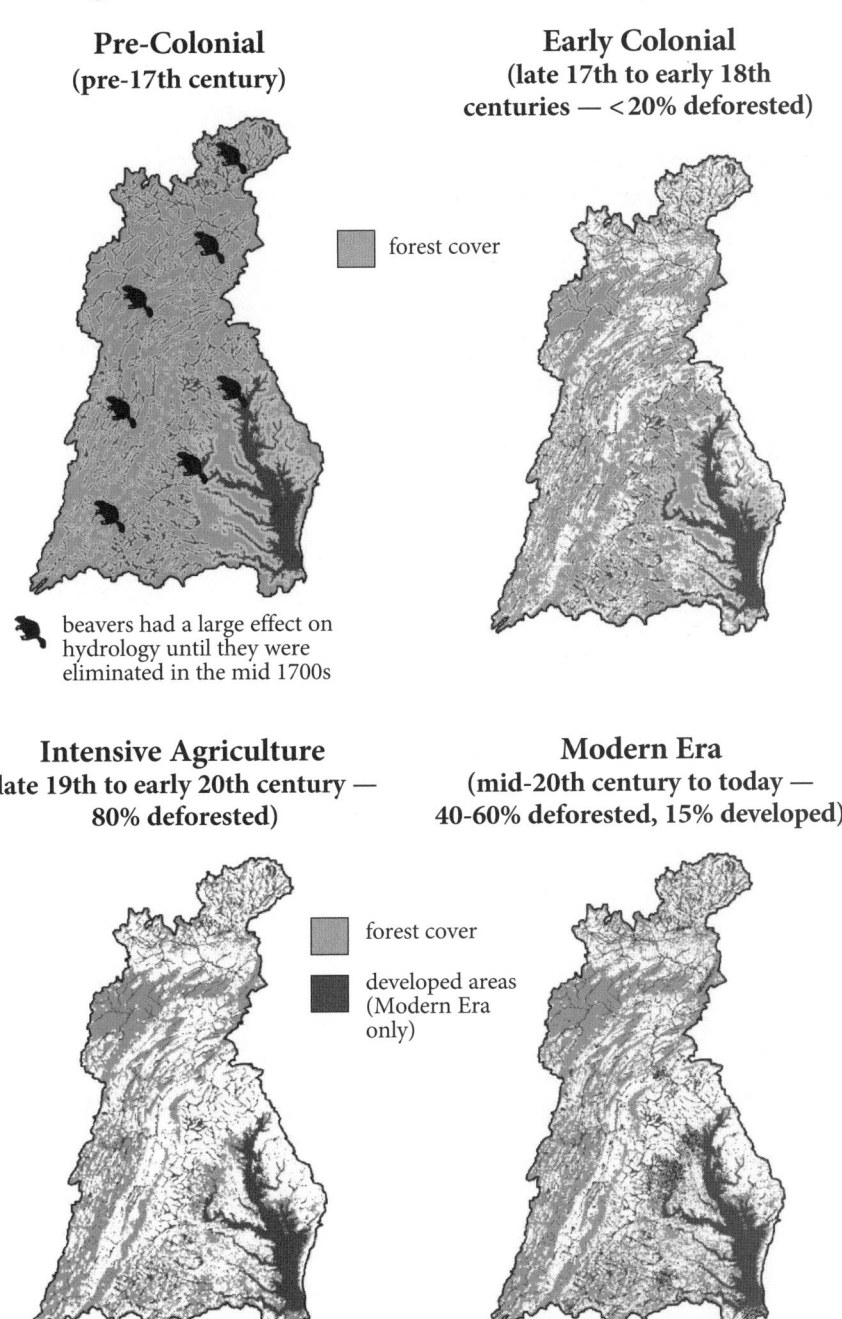

Pre-Colonial
(pre-17th century)

forest cover

beavers had a large effect on hydrology until they were eliminated in the mid 1700s

Early Colonial
(late 17th to early 18th centuries — < 20% deforested)

Intensive Agriculture
(late 19th to early 20th century — 80% deforested)

Modern Era
(mid-20th century to today — 40-60% deforested, 15% developed)

forest cover

developed areas (Modern Era only)

clay particles are fertilizers and other chemicals applied either to the land or discharged directly into the water.

In addition, pollen grains, seeds, and other parts of plants including twigs, leaves, and pieces of bark are carried to and deposited in the estuary. The distance materials are transported before being deposited is related to the size and weight of the exported particles and the depth and turbulence of the receiving water. Materials originating in the atmosphere also reach the estuary. These include the wind-blown pollen of many plants as well as fertilizers such as nitrogen and phosphorus, and chemicals discharged from industrial sources and automobiles. Along with these extraneous items, organisms that live in the estuary itself, represented by the outer coverings (frustules) of diatoms, spores of other microscopic organisms such as dinoflagellates, parts of insects, invertebrates, and fish scales settle to the bottom and are buried by the accumulating sediment. Only those organisms or parts of organisms resistant to physical or biological decay, and chemicals that are more or less inert, are preserved. Despite this caveat, the sediments are a rich repository of the history of the Chesapeake.

Paleoecology

Our approach to understanding the ecological evolution of the Chesapeake Bay is through the science of paleoecology, which begins with collecting cores of sediment from areas where sediment is deposited. The organisms and chemical substances preserved in a core are a record of what was present or happening over time. Thus a sediment core can be likened to an ancient manuscript of hieroglyphics that must be deciphered and translated into a language we can understand. We can then "read" each section of the core as though reading pages in a book that describes events, which took place hundreds to thousands of years ago.

Collecting and Processing Sediment Cores

The first task is to locate undisturbed areas of continuous rapid sedimentation, which provide a high time resolution. Such areas can be identified by simply sticking a long pole into the sediment at various locations at a site. The amount of sediment deposited at a particular site can vary greatly within the space of a few meters. Many bottom sediments are disturbed by natural events such as storms and burrowing organisms, and by human impacts, for example dredging. Information about areas disturbed by humans is often available from historical records of dredged areas and from various geologic studies and environmental reports.

To ensure retrieval of an undisturbed core, or one that has undergone the fewest disturbances over time, we collect 6 to 12 cores at a location. This number, based on comparing the diameter of the core (2.5 cm) with the size and frequency of disturbed areas evidenced from maps of surface sediments in the Bay, estimates that the probability of collecting an undisturbed core in the fresher parts of the Bay is one in six, whereas in the more saline parts where burrowing organisms and other disturbances are more frequent, we

Figure 3. William Hilgartner (right) with two undergraduate students collecting a sediment core from a boat in a Chesapeake Bay tributary. Photograph courtesy of Grace Brush.

would need to collect 12 cores to ensure collecting at least one that is undisturbed. This rule of thumb has served us well. Our initial coring sites in the 1980s were located with the help of the late Dr. Owen Bricker, a geochemist with the Maryland Geological Survey and later with the U.S. Geological Survey, who was doing geochemical analyses of pore waters in the sediments to address questions related to toxic contaminants in the Bay.

Once the location is established, a cylindrical core outfitted with a piston (Figure 3) is lowered into the sediment by hand or with a power tool. The piston creates a vacuum, which prevents the core of sediment from dropping out of the tube as it is being extracted from the site. Hand-driven cores are generally one meter in length, although we have retrieved cores two meters long. Generally, we collect marsh cores using a vibrocorer, a metal cylinder with a jackhammer type of device powered by a small motor, which drives the cylinder through the marsh deposit without compacting the sediment. The top and bottom of the core are capped and the top labeled.

In the laboratory, we extrude the core of sediment from the cylinder, divide the core in half lengthwise, so as to see clearly the layers of sediment (Figure 4). Thus, we can examine the macroscopic stratigraphy to determine whether or not the strata are intact. In some cases, sections of a core may be separated by zones of disturbed sediment — if we date these sediments, we can determine the period when the disturbance took place. The core is then sectioned at 0.5- to 2.0-cm intervals: each sample is placed in a plastic bag, labeled and stored at 4°C until processed for analyses of the "indicators."

Different processes are employed to isolate pollen, seeds, diatoms, etc., as well as carbon, phosphorus, nitrogen, metals, and isotopes of carbon and nitrogen and other elements. For example, seeds are isolated from the sediment by washing the sediment with water through a series of sieves and microscopically picking the seeds out of the residue on the sieves. Pollen, on the other hand, requires multiple washes of various chemicals to remove materials other than pollen from the sediment. Diatoms also require multiple washes with different chemicals than those used for pollen to isolate the silica shells (frustules).

Figure 4. Grace Brush observing the stratigraphy of a sediment core that has been split longitudinally. Photograph courtesy of Johns Hopkins University.

Dating the Cores

Several methods are available for dating cores, and we generally use more than one method for an individual core.

- Carbon-14, with a half-life of 5,730 years, is used to date sediments older than 500 years and younger than 50,000 years.
- Lead 210, with a half-life of 22 years, allows dating sediments 75 to 100 years or younger.
- Peaks of Cesium 137 in sedimentary horizons represent atomic testing done in 1957 and 1963.

- Presence of volcanic ash, pollen, and seeds of introduced plants or the sudden absence of pollen and seeds of known plant extinctions can be correlated with historic events.

In dating Chesapeake sediments deposited since European settlement, we use the decrease in oak pollen and increase in ragweed pollen associated with historically dated deforestation and agriculture and the absence of chestnut pollen following the extinction of chestnut trees in the 1930s. Sediments that predate European settlement are Carbon-14 dated. (See Figure 5.)

Developing a Chronology of Events

A second preliminary task in paleoecological studies is to refine the dating of individual cores in order to construct the actual sequence of events or chronologies. The hydrodynamics of the estuary are such that sedimentation varies a great deal from day to day, month to month, year to year. Sedimentation also varies spatially, so that a meter of sediment at one location can span a thousand years or more, while at another location in near proximity a few hundred years or less. Average sedimentation rates between dated horizons in individual cores vary enormously, whether the sediment is pre- or post-Colonial.

In order to construct a chronology for an individual core, we calculate a sedimentation rate for each sample of the core and adjust that rate to years. These calculations must be done for each core regardless of how spatially close they are, because of the high variability in sediment accumulation within a few meters. This information is essential for reconstructing changes in populations of organisms, sediment, and chemicals within each core; for example, ten diatoms in a cubic centimeter of sediment provides a quite different population estimate if that cubic centimeter was deposited in one year or a hundred years.

In developing the chronology of a core, we first consider the hydrodynamics of silt and pollen grains. Silt constitutes over 90% of the sediment in

---

Figure 5 (opposite page). Schematic drawing of a sediment core showing stratigraphic indicators related to dated historical events. Illustration by Sandy Rodgers.

Dating a Chesapeake Bay Sediment Core

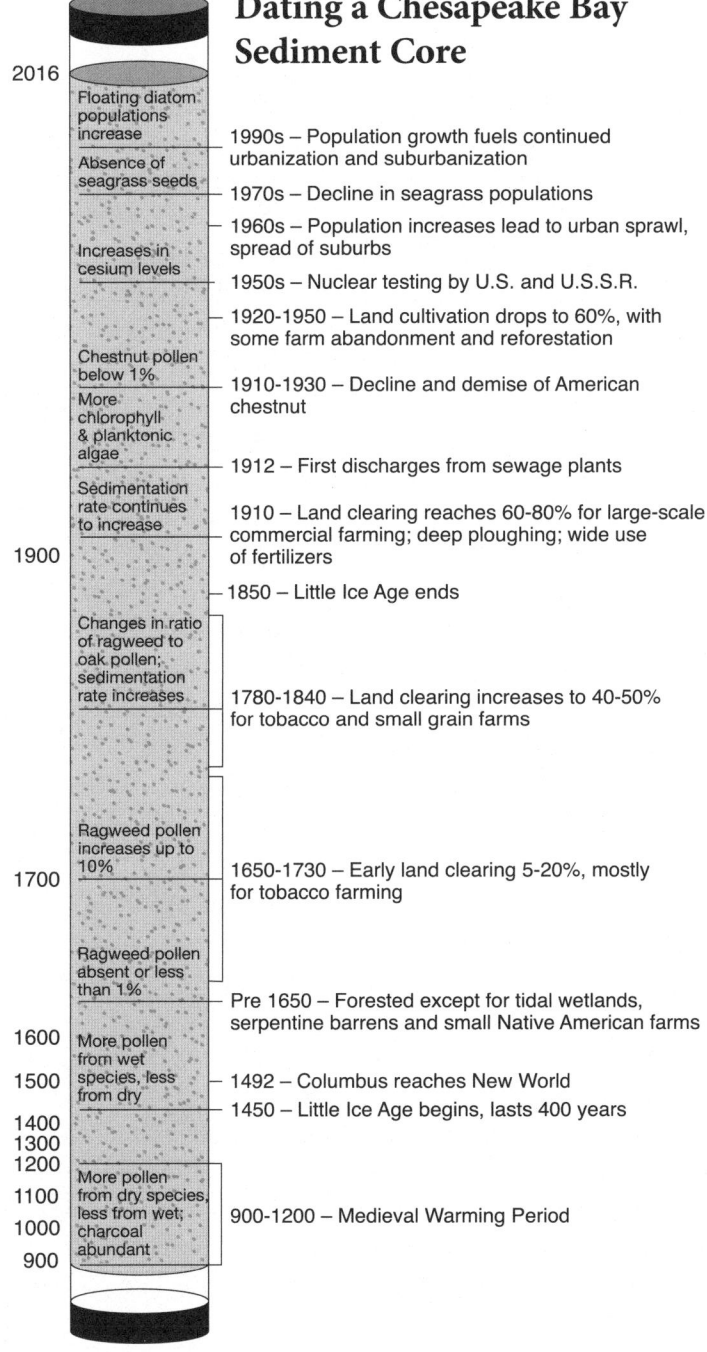

2016

Floating diatom populations increase

Absence of seagrass seeds

Increases in cesium levels

Chestnut pollen below 1%

More chlorophyll & planktonic algae

Sedimentation rate continues to increase

1900

Changes in ratio of ragweed to oak pollen; sedimentation rate increases

Ragweed pollen increases up to 10%

1700

Ragweed pollen absent or less than 1%

1600

1500
More pollen from wet species, less from dry

1400
1300
1200
1100
More pollen from dry species, less from wet; charcoal abundant
1000

900

1990s – Population growth fuels continued urbanization and suburbanization

1970s – Decline in seagrass populations

1960s – Population increases lead to urban sprawl, spread of suburbs

1950s – Nuclear testing by U.S. and U.S.S.R.

1920-1950 – Land cultivation drops to 60%, with some farm abandonment and reforestation

1910-1930 – Decline and demise of American chestnut

1912 – First discharges from sewage plants

1910 – Land clearing reaches 60-80% for large-scale commercial farming; deep ploughing; wide use of fertilizers

1850 – Little Ice Age ends

1780-1840 – Land clearing increases to 40-50% for tobacco and small grain farms

1650-1730 – Early land clearing 5-20%, mostly for tobacco farming

Pre 1650 – Forested except for tidal wetlands, serpentine barrens and small Native American farms

1492 – Columbus reaches New World

1450 – Little Ice Age begins, lasts 400 years

900-1200 – Medieval Warming Period

much of the Bay and pollen grains represent primarily plants growing around and in the Bay. Based on their physical properties, both silt and pollen are expected to behave similarly in water or air transport: the settling velocities in water and air are calculated using the Stokes equation for the fall of small particles (McNown et al. 1951). To determine if pollen grains behave similarly in a fluid, we measured settling velocities of individual grains in lab experiments and compared them with settling velocities derived from the Stokes equation. They were statistically equivalent. We further assumed that if silt particles were entering the Bay primarily through stream transport and pollen particles mainly from the atmosphere, we were dealing with two populations of particles that enter the estuary through different media but would behave similarly during transport and deposition.

To test the assumption that the sources differed, we measured amounts of suspended sediment and pollen seasonally at locations throughout the length of the Bay. The data indicated that introduction of pollen into the Chesapeake was at its maximum during pollination in the spring, whereas the influx of suspended sediment was fairly uniform, varying with runoff throughout the year. These findings supported our assumption that silt was being carried by river runoff while pollen was transported to the estuary primarily by air. Therefore, if the ratio of pollen to silt/clay is low, the deposition of sediment would be high and vice versa: if the pollen to silt/clay ratio is high, the deposition of sediment would be low. Using techniques described in Brush 1989, we can calculate the sedimentation rate for each individual sample, determine the number of years represented by each sample, and with that information construct a chronology for an individual core.

Estuarine Paleoecology

During the 1970s concern over deterioration of the Chesapeake Bay's aquatic resources was gathering strength, particularly because of the widespread disappearance of submerged aquatic vegetation (SAV) and the decline of commercial fisheries. What were the effects on these resources of runoff from the land, sewage effluents, and toxic chemicals? Mounting public demand for action was instrumental in a $27-million appropriation from Congress, administered through the U.S. Environmental Protection Agency (USEPA), to "restore the Bay."

First Paleoecological Research in
Chesapeake Bay Focused on SAV

Until the Chesapeake Bay Program got underway, only a few paleoecological studies had been done in the Bay, and these were from sediment cores collected at the mouth of the estuary, which gave a very general postglacial history of vegetation. Extensive paleoecological research had been done in the Great Lakes, however: in order to determine the timing and causes of eutrophication, researchers there analyzed diatoms and phosphorus preserved in sediment cores. David Flemer, an EPA scientist working in the Chesapeake region had previously worked in the Great Lakes and knew how paleoecological studies had contributed to an understanding of eutrophication in those systems. While he strongly supported the potential for such research, some were skeptical about whether undisturbed cores could be retrieved from tidal systems such as the Chesapeake; however, physical scientists working on Bay hydrodynamics predicted that the Chesapeake environment is similar in many ways to a large settling basin and thus should be suitable for paleoecological studies.

Eventually, I was asked to submit a proposal to the EPA for a paleoecological study that would focus primarily on the history of SAV. Although I had access to a laboratory with a fume hood and microscopes from previous grants, the EPA award was for only one year, little enough time to research core locations, let alone recruit students.

Because Owen Bricker had located sites for the biogeochemical studies he was undertaking, we decided that these would serve as ideal sites for our paleoecological research. I immediately contacted students I knew or learned about from colleagues. Frank Davis and Sherri Cooper were the first Ph.D. students to work on the project. Both had strong biological/environmental backgrounds, and while neither had any experience in paleoecology, they were interested and innovative. Suzanne Bricker, then an undergraduate, worked as a research assistant and would later pursue a Ph.D. in oceanography at the University of Rhode Island. They were assisted by undergraduates who helped collect, sample, and process the cores — time-consuming arduous tasks.

While a major focus of what became the Chesapeake Bay Program was restoration of SAV — critical habitat for many estuarine organisms and a determinant of healthy water — it was difficult to design a recovery plan without a relatively accurate knowledge of the history of SAV loss and whether such events had occurred previously. Some researchers suggested that living populations go through boom-and-bust cycles, which might explain recent declines. I believed that by analyzing SAV seeds preserved in the sediment, we could determine whether the present losses were either unique or cyclical, and, if the former, when the deterioration began.

Landscapes and the Bay

The Pre-Colonial Landscape

In reconstructing the history of plants from fossil pollen analyses, we first determined how the present vegetation cover is represented by pollen, as some plants produce much more pollen than others. For example, oak and pine trees are prolific pollen producers; insect-pollinated species are less well represented in the sediment; and the pollen of other trees such as tulip poplar is rarely seen in sediments, either due to low pollen production or poor preservation. Ragweed produces large amounts of pollen compared with corn, for example. We established these tree-pollen relationships by comparing the percentage of the different pollen types in surface sediments of the Potomac River with the percentage of tree species in the adjacent vegetation obtained from data collected for the vegetation map which we constructed in 1978 (Brush et al. 1980, Brush and DeFries 1981). We chose the Potomac River for this comparison as we had an ongoing study of sedimentation rates throughout the river based on pollen and hence had data on pollen in surface sediments. Similarly, we compared the density of species in SAV beds with the density of SAV seeds in proximity to the beds. It is also necessary to approximate the area of vegetation represented by pollen and seeds (Figures 6 and 7) preserved in sediment, since both of these entities reach the sediment by atmospheric and fluvial transport and can occur in sediment some distance from the parent plants. To do this comparison, we examined pollen and seeds in sediment cores with vegetation along transects adjacent to the coring sites. In addition to the field studies, we also conducted field and laboratory (flume) studies to determine dispersal distances of pollen and seeds (Brush and Brush 1972, 1994; Davis 1985; Hilgartner and Brush 2006).

> ### Sediment Collection Sites
>
> - Indian Creek floodplain of the Anacostia River
> - Jug Bay in the Patuxent River
> - St. Mary's River off the lower Potomac River
> - Otter Point Creek in the Bush River
> - Monie Bay bordering the Nanticoke River
> - Furnace Bay in the upper Chesapeake
> - Magothy and Nanticoke Rivers

Sediment cores collected in a variety of depositional settings including the estuary, marshes, and floodplains provided a history of pre-Colonial Chesapeake vegetation (Figure 8). Two of the 18 sediment cores recovered from the floodplain of Indian Creek, a tributary of the Anacostia River close to Washington, D.C., contain a complete Holocene (post-glacial) record; one core spans 14,500 years and a second contains a 10,000-year record. The pollen and seed records show the evolution of vegetation after the last glacial period from a spruce-and-fir forest to a warmer but wet landscape that supported stands of hemlock, followed by the modern oak forest 6,000 years ago. However, pollen preserved in cores from St. Mary's River off the lower Potomac and the Magothy River in the mid Chesapeake (all of which spanned thousands of years) show different pollen sequences as climate changed during post-glacial time.

- **Anacostia site.** The substrate is clay/loam and the present vegetation is dominated by tulip poplar, river birch, and sycamore; however, 6,000 years ago the vegetation was changing from hemlock to primarily oak and shrubs of the blueberry family (Yuan 1995) (Figure 9).

- **St. Mary's site in the lower Potomac.** Vegetation is dominated by willow oak and loblolly pine on a clay and sandy loam; about 6,000 years ago, it was shifting from Cupressaceae, a family that includes Atlantic white cedar, bald cypress, and red cedar (the pollen of which are indistinguishable) to primarily sphagnum and alder (Brush, unpublished data).

- **Upper Magothy.** Vegetation is dominated by chestnut oak, post oak, and blackjack oak along with tulip poplar on sand and fragipan; between 4,000 and 5,000 years ago, the vegetation was mainly sweet gum and black gum (Brush, 1986).

The response of vegetation to climate change was synchronous at all locations but the species differed depending on the geology/soils substrate. Climate change is superimposed on a geologic and topographic template put in place in this area millions of years ago.

Evidence of pre- and post-Colonial changes is recorded in paleoecological studies of many marshes throughout the Chesapeake; these changes include the following.

- **Medieval Warm Period** (~350-1250 A.D.). Of the cores we have analyzed, Jug Bay's marsh deposits on the Patuxent River are where we found evidence of this warming period, which is characterized by abrupt decreases in the pollen and seeds of wet species such as walnut and wild rice and large increases in drier species, including American holly (Khan 1993) (Figure 10).

- **Changes in sea level.** Pre-Colonial changes are recorded in shifts in SAV in sediment cores collected in the brackish Monie Bay in the Nanticoke River (Thornton 1992).

- **Changes in tidal waters.** Tidal freshwater areas in the Otter Point Creek marsh in the Bush River and Furnace Bay in the upper Chesapeake were subtidal up until colonization (Hilgartner 1995, Pasternack 1998).

The Post-Colonial Landscape

From Forests to Farms and Freshwater Marshes

The pre-Colonial forested land, governed primarily by climate, supported an assemblage of species different from the one that occupies the land now dominated by humans. The photosynthetic organisms that provide sustenance for the Bay's watershed ecosystem changed from predominantly trees and shrubs to mostly grasses and forbs. As the plants changed, the

Tree Pollen

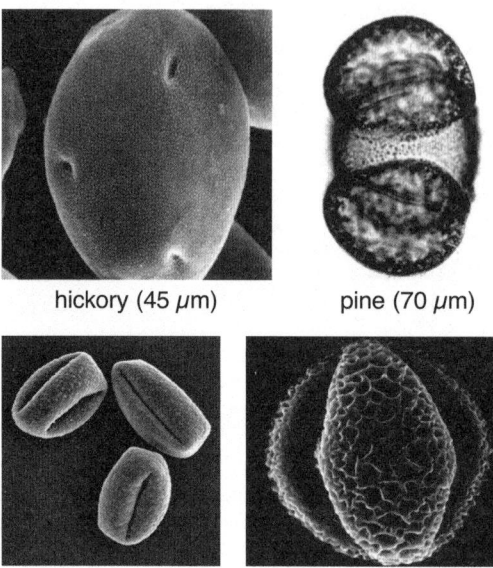

hickory (45 μm)   pine (70 μm)

oak (21 μm)   willow (14 μm)

Figure 6. Fossil tree pollen embedded in layers of sediment provides a means for dating core segments. Micrographs by Grace Brush.

Tree Macrofossils

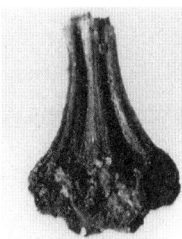
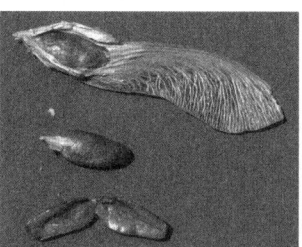

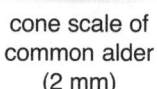

cone scale of
common alder
(2 mm)

fruit and seeds of maple
(seeds are 12 mm long)

Figure 7. Macrofossils like these are also used for reconstructing environmental conditions. Micrograph of alder by William Hilgartner.

land changed and relationships between land and water were fundamentally altered.

In contrast to the Native Americans who generally did not own fixed properties but were more mobile depending on the season, the first colonial settlers claimed tracts of land as personal property: in the eastern part of the country, individual properties were identified by a system called metes and bounds. The resulting irregular property boundaries were marked by "witness trees" that grew along property lines.

Early settlers used hoes and axes to clear small patches of forest within their property to plant crops. Plots were cultivated for four to five years and then abandoned for as long as twenty years during which time new plots were cleared and cultivated. Known as fallow farming, this practice allowed time for natural restoration of nutrients in the soil. During this period, only 3 to 20% of the land was under cultivation at any one time. Thus, soil erosion was minimal. Cultivated plots were buffered by surrounding wooded areas and much of the mobilized sediment was deposited in the vicinity of the plots.

As populations increased, crop rotation replaced fallow farming, which allowed continuous cultivation of farmed plots and the cultivation of more land at any one time. The wooden plow was replaced by the iron plow and later by the steel plow, which dug even deeper into the soil. Still, the farms were small and surrounded by woodlots and hedgerows; soil loosened by cultivation for the most part remained on the land and adjacent floodplains. However, by the late 19th century, large expanses of land were farmed with very few buffers; eroded soil was transported to the tributaries, which became clogged with sediment. There was much concern about the loss of topsoil as well as infilling of the tributaries, which curtailed navigation, a principal avenue for trade (Gottschalk 1945).

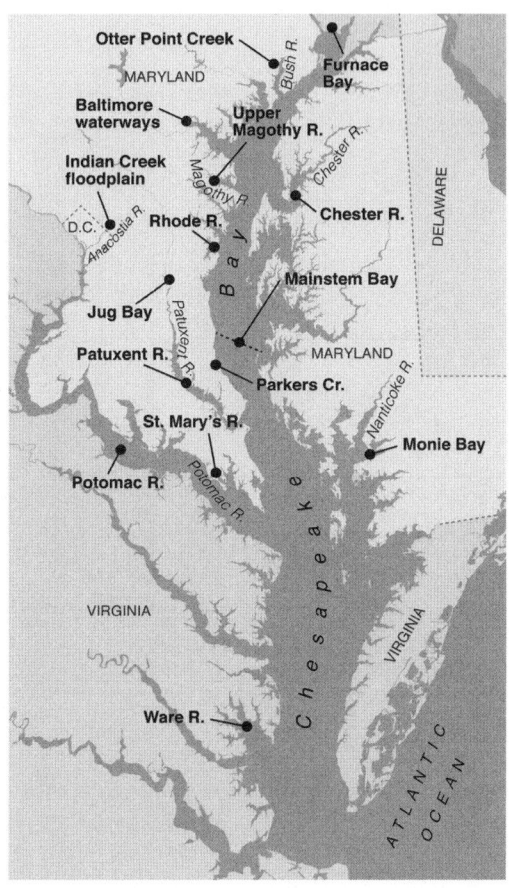

Figure 8. Locations of sampling sites in the Chesapeake Bay and in tributaries referred to in the text. Multiple cores were taken at these sites: Otter Point Creek (5), Baltimore-area waterways (14), Furnace Bay (2), Upper Magothy (3), Indian Creek floodplain (18), Mainstem Bay (4), Patuxent River (3, including Jug Bay), Potomac River (15), Monie Bay (3), and Ware River (3). Map from Vectorstock.com with sampling sites from Grace Brush added by Sandy Rodgers.

The first European settlers in the early 1700s were not unaware of the physical mosaic of the land, choosing individual properties to farm with

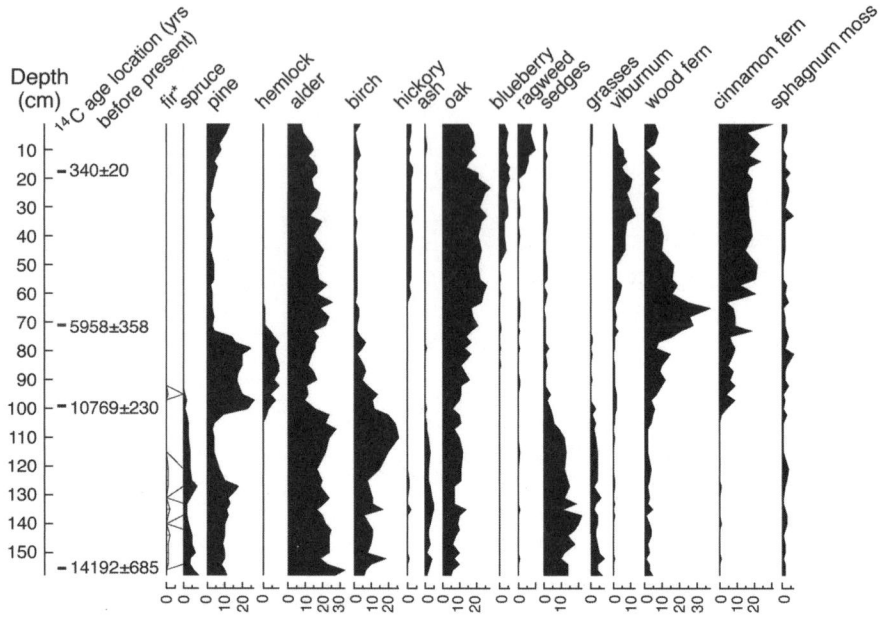

Pollen percentages
(*Symbols in fir profile represent a 10x exaggeration)

Figure 9. A 14,000-year pollen profile from a sediment core collected in the Anacostia River floodplain shows how concentrations of plant species varied over the millennia. Graph redrawn from Brush (1984).

good soil, southern slopes, and flat topography (Bain and Brush 2008). Northern, less sunny slopes remained forested. Hence, land cleared for farming coincided more or less with the natural mosaic. This pattern of finding the best landscape features conducive for farming has been shown elsewhere (e.g., Helms 2000). In many cases, tree species were used to identify the most productive soil. Interestingly, at least in the Gwynns Falls watershed, the areas originally designated as farm properties were on the ridges rather than the floodplains, which would have had to be drained (Bain and Brush 2008).

Towards the end of the 19th century and up until the 1930s, almost all land throughout the Chesapeake watershed was farmed, regardless of topography or substrate. Although the specifics of farming differed, the changes from forest to farmland were the same. The legacy of early agricul-

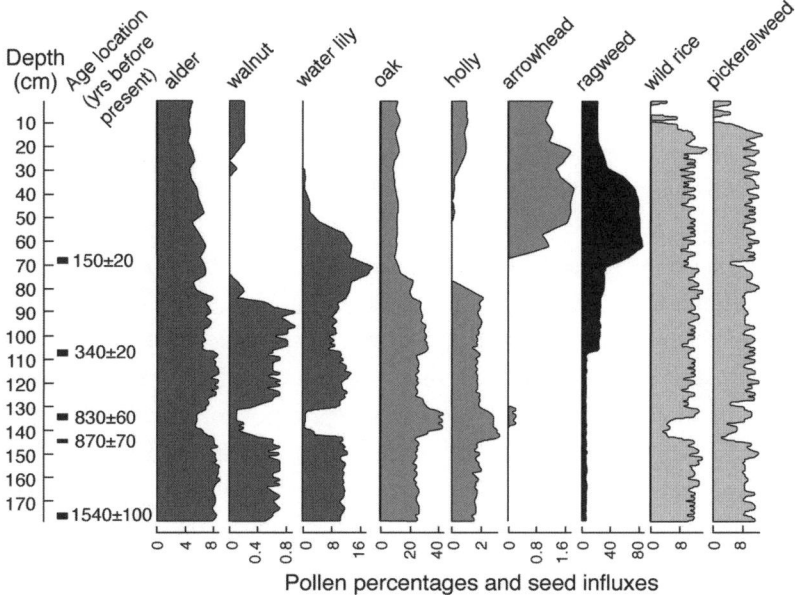

Figure 10. Pollen and seed profiles from Jug Bay in the Patuxent River showing a decrease in wet species and an increase in dry species during the Medieval Warm Period (830-870 years ago). The profiles also reveal the increase in concentrations of ragweed since the Colonial period. Graph from Grace Brush (unpublished data).

ture patterns can still be seen in some areas that remained agricultural, but in urban areas the original mosaic has been erased and superseded by a mosaic based on activities unrelated to the landscape, except for large highways which were built later and followed topographically low areas (Bain and Brush 2005). Other land uses including mining for iron and chromium and quarrying for marble — substances related to particular geologies — coincided with the mosaic template.

When trees were cleared for agriculture, rain no longer soaked into a forest floor with deep litter but rather flowed off the land, carrying with it soil that eventually led to the formation of new land forms. At the mouths of tributaries, where flow rates lessened, sediment was deposited in delta formations, which became freshwater marshes. Later, when more than 60% of the land was deforested, the larger sediment loads were transported and deposited throughout the estuary.

Marsh development is related to sedimentation, topographic elevations, tides, and climate factors including precipitation and wind, leading not only to high and low marshes but also to intricate patterns of sediment accumulation or non-accumulation within the marshes. Indeed, the diverse habitats within the marsh environment result from those dynamics. In studying patterns of sedimentation within the Otter Point Creek Marsh next to the Bush River, Greg Pasternack placed tiles throughout the marsh, and measured sediment accumulation (or lack of accumulation) on the tiles biweekly over a year, a method similar to the Surface Elevation Tables (SET) used by other researchers to measure marsh accretion. He showed that by taking into account historical sedimentation rates reconstructed from sediment cores, a diffusion model provided a good approximation of tidal marsh growth (Pasternack and Brush 2002). Pasternack and Linda Hinnov also showed that water levels in Otter Point Creek responded rapidly to weather change, which accounted for rapid changes in sedimentation (Pasternack and Hinnov 2003).

Other changes were also taking place on the land. As marshes were growing in the tidal freshwaters, wet areas once created in inland streams by beavers began diminishing as these animals were hunted for their fur. Floodplains and soggy wet lowlands were drained for agricultural crops. Areas behind dams built to power grist mills were inundated but the areas below the dams became drier. The drying of the land reduced its anaerobic potential so that atmospheric nitrogen, transformed microbially and photosynthetically to organic nitrogen compounds, could not be denitrified to elemental nitrogen and released harmlessly to the atmosphere. Thus nitrogen, through many microbial transformations, remained on the land providing an unlimited supply of nutrients: in runoff, this soil-bound nitrogen was transported into the estuary where it fueled the growth of estuarine primary producers. When algae, which make up many of these primary producers, grow beyond their potential for being grazed (for example, by oysters, vast populations of which were being overharvested), the cells sink to the bottom waters where they decompose, using much of the dissolved oxygen in those waters. This process can lead to severe depletion of oxygen concentrations, called hypoxia, or to the complete depletion of oxygen, anoxia. Meanwhile, in order to increase agricultural production, nitrogen — first mined from guano and saltpeter and later chemically synthesized — and phosphorus were added as fertilizers, thus exacerbating still more the potential for increased estuarine algal production and eventual eutrophication and anoxia.

As forests were cleared throughout the watershed and replaced not only by farms but increasingly by impervious surfaces such as buildings and roads, greater volumes of fresh water were running into the estuary. Organisms had to adapt again to new conditions of salinity, as some estuarine waters became less saline. Thus, many brackish plant species, such as *Ruppia* in the upper parts of the estuaries, migrated into the lower estuaries where salinity was higher and were replaced in the upper estuaries by freshwater plants such as *Vallisneria* (Figure 11).

Agriculture and trade brought many species from far off places. Some, like Japanese honeysuckle (*Lonicera japonica*) and dandelion (*Taraxacum officinale*), spread rapidly and outcompeted native species. However, a species native to North America, the common ragweed (*Ambrosia artemiisifolia*), has emerged as the most invasive species on Earth. Mike Martin conducted research on the genetics of ragweed and its spread through the landscape. After investigating the mechanisms of pollen release and transport from common ragweed, which explained the plant's ability to spread by cross-pollination, Martin studied the genetics of extant ragweed populations from east of the Mississippi, south to Florida, and north to Nova Scotia, and described the geographic distribution of different genotypes of existing populations (Martin et al. 2011). He then studied the genetics of ragweed-seed populations collected a hundred years ago from the eastern

Underwater Grass Seeds

 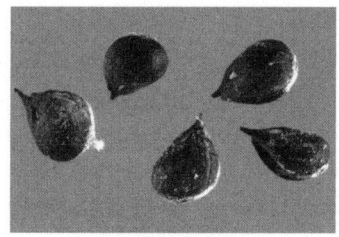 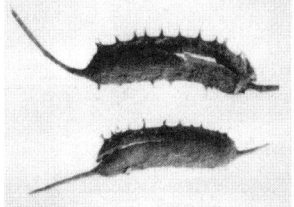

Vallisneria americana (wild celery) (2.4 mm)

Ruppia maritima (ruppia) (2.2 mm)

Zannichellia palustris (horned pondweed) (2.5 mm)

Figure 11. The concentrations of underwater grass (submerged aquatic vegetation seeds in segments of cores are important indicators of the health of bottom waters. Micrographs by William Hilgartner.

U.S. and Canada, which were stored in herbaria (plant museums). He reconstructed a geographic phylogeny of the genotypes of common ragweed, which is related somewhat to land use and which explains, at least partially, the enormous success of this plant as an invasive (Martin et al. 2014). As the land was changing, some plants were developing the wherewithal to adapt to a new environment.

Paleoecological Record of Changes on the Land

The changes from forested to deforested landscapes that accompanied European colonization are recorded in pollen records from sediment cores collected in marshes, tributaries, and the estuary by large decreases in oak pollen and correspondingly large increases in ragweed pollen. As deforestation proceeded, native ragweed invaded disturbed areas in proportion to the amount of land cleared. Ratios of tree pollen to ragweed pollen to non-ragweed herbaceous pollen record the history of deforestation at different localities (Figure 12). Prior to European colonization, ragweed was either absent or made up 1% of the total pollen in a sample. When 20% of the land was cleared, the proportion of ragweed pollen ranged between 1% and 10%, but was generally 5%; when 40% of land was cleared, ragweed increased to >20%. Records of times and the extent of forest clearance are available in historical documents and, since 1840, from the U.S. Agricultural Census records. The records are used to date the ragweed/oak ratios that record the periods of forest clearance in the sediment cores.

Different patterns of sedimentation related to rapid changes in land use (recorded in the oak:ragweed pollen ratio) resulted in the evolution of a very diverse ecosystem of deltas and freshwater marshes in the tidal tributaries. Pollen and seeds of terrestrial and aquatic plants in sediment cores collected from marshes record the evolution from subtidal to various wetland and forested habitats. William Hilgartner's (1995) description of the temporal and spatial evolution of marsh habitats at Otter Point Creek in the Bush River shows the effect of deforestation and road building on the evolving delta marshland. In 1940, U.S. Route 40 transected a part of the marsh, which caused localized flooding, creating a forested wetland. In 1970, the completed Route 24, which intersected Route 40, resulted in flooding of the previously formed forested wetland, converting it back to a marshland. Martha Jarosewich (1985) traced changing shorelines and delta development at Furnace Bay and at the mouth of the Elk River using historical maps

The Rise of Ragweed

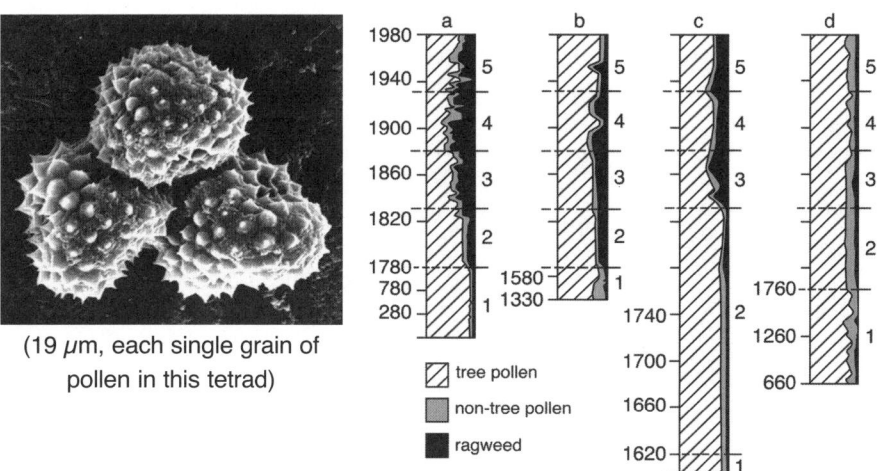

(19 μm, each single grain of pollen in this tetrad)

tree pollen

non-tree pollen

ragweed

Figure 12. A micrograph of a tetrad of ragweed pollen (left). A graph of pollen profiles (right) shows different land use periods using the ratios of tree pollen to ragweed pollen to non-tree pollen excluding ragweed pollen at four different locations. (a) Magothy River, (b) Furnace Bay, (c) Potomac River, and (d) Nanticoke River. 1, 2, 3, 4, and 5 refer to different periods of land use: (1) pre-Colonial, (2) >20% of land in cultivation, (3) 40-60% of land in cultivation, (4) period of maximum deforestation, and (5) farm abandonment and afforestation. The left axis is years AD. Ragweed pollen increased in proportion to the amount of land deforested. The site on the Nanticoke River (d) experienced minimal land clearance in post-Colonial time. Micrograph from Grace Brush; graph redrawn from Brush and Brush (1994).

and sedimentological data from dated cores. Humaira Khan (1993) used sediment cores to track the development of high and low marshes at the fresh tidal Jug Bay in the Patuxent River. She showed the change from a low to a high marsh resulting from the building of a railroad across the marsh in 1895 for the transport of tobacco from the inland to the estuary. The railroad embankment reduced flooding from daily to bimonthly in the section of the marsh transected by the railroad. Khan's studies also showed how low and high marshes differed with respect to their ability to buffer the estuary, with high marsh woody plants decomposing at a much slower rate than the non-woody low marsh plants and hence retaining nutrients for a longer period of time.

The Chesapeake Bay Ecosystem

The Pre-Colonial Estuary

Archeological and historical records indicate that the late pre-Colonial estuary contained a highly productive food web. In an early-17th-century account of the Chesapeake, Captain John Smith wrote that "oysters lay as thick as stones," while fish populations included "sturgeon, grampus, porpoise, seals, stingray, brits, millets, white salmon [rockfish], trout, soles, [and] perch of three sorts." There is an apocryphal story that onboard ship he and his men were so surrounded by schools of fish that they tried to catch them with frying pans! Miller (1986) records archeological data from both low- and high-salinity areas that show catfish and white perch were most abundant in areas low in salinity along with striped bass, long nosed gar, and sturgeon, which occurred commonly. In the high-salinity areas, sheepshead, black drum, and white perch were most common, with some striped bass and sturgeon. Blue crab and oysters were abundant at most sites. Oyster reefs, common throughout the Bay, were considered navigational hazards; shell size of oysters was much larger in the early 17th than in the later 17th century, when oysters began to be harvested as a crop. Miller (1986) suggested that fishing probably focused on benthic habitats in shallow waters because hooks and lines were the most common fishing equipment. Pelagic species, such as bluefish and sea trout, spot, and Atlantic croaker were not found at the archeological sites. Some saltwater species were found in areas that are now less saline, indicating changing salinities in the Bay with land use. Historical records describe the large harvests not only of oysters but also of shad and striped bass, which spend much of their lives in higher-salinity waters but migrate to fresh water in the upper tributaries to spawn (Cronon 1986).

The paleoecological record of organisms shows that up until the period of colonization, the Bay was largely a benthic system with a continuous record of SAV and a high diversity of predominantly benthic diatoms as well as an abundance of detritivores. Foraminiferal populations indicate a saltier Bay than in post-Colonial time when deforestation resulted in higher freshwater discharge from the land (Sowers and Brush 2014).

The Post-Colonial Estuary — From a Benthic to a Pelagic System

As the land in the watershed was transformed from forests to farms and large areas of impervious ground, the estuary itself underwent major ecological changes. At the same time that a greater volume of soil and nutrients was being deposited into fresh waters, oyster harvests and the dismantling of oyster reefs were expanding significantly.

In addition, commercial fishing intensified, even as new dams were being built, cutting off the freshwater spawning grounds of important species such as shad. The Chesapeake Bay, before Colonial settlement, had been dominated largely by benthic processes: bottom waters were oxygenated and SAV (important for water clarity and water quality) appeared to thrive. The "perfect storm" of increasing soil runoff, nutrient loading, benthic harvesting of oysters and bottom fish, and dam construction were key factors in the Bay's transition from a benthic-dominated ecosystem to a pelagic ecosystem. (See Figures 13 and 14.)

Land Use and Sedimentation Rates

Following Colonial settlement, sedimentation in the estuary and tributaries increased in proportion to the degree of forest clearance; within individual tributaries, sedimentation was highest in the upper reaches, and lower at the mouths of the tributaries. Ruth DeFries studied the history of sedimentation patterns in the Potomac River based on pollen analyses in 20 sediment cores collected throughout the river. Her findings showed that sedimentation rates from 1634 (initial European settlement) to 1980 ranged from 0.2 to 1.7 cm per year with the highest rates in the upper estuary and lower rates at the mouth. Only 15% of the fine-grained sediment entering the head of the subestuary Port Tobacco was transported into the Potomac River (DeFries 1980).

Fred Scatena related sedimentation rates to land use in the Anacostia

River. In the period between initial European settlement and the founding of Washington, D.C., in 1791, the major sources of sediment were farm fields and roadbeds. However, as the area became urbanized, about 80% of the sediment load came from a much smaller land area, including quarries and construction sites or stream channels rather than agricultural fields. Most of the sediment from urban land was deposited as colluvium on floodplains and in surface-water impoundments. A small fraction was transported to the tidal embayment at the mouth of the Anacostia (Scatena 1987).

Dan Bain calculated rates of sedimentation in floodplains by measuring chromium profiles in sediment cores collected in a floodplain in the Gwynns Falls Watershed: chromite was mined from 1820 to 1880 in Soldiers Delight, which is drained by a tributary to the Gwynns Falls. Chromite mining occurred during a period of rapid deforestation for agriculture. This study showed that sedimentation rates before 1900 ranged from 0.45 to 1.2 cm per year in the floodplain near the mining activity. These rates are higher than the post-1900 sedimentation rates of 0.08 to 0.46 cm per year, indicating that widespread agricultural clearing may have occurred in the mid-1850s, rather than toward the end of the 19th century in the upper Gwynns Falls watershed (Bain 2003, Bain and Brush 2005). Peak sedimentation rates between 1840 and 1880 were also shown at Otter Point Creek (Hilgartner and Brush 2006).

Taken together, these studies show a clear relationship between sedimentation and local land use. As small farms (local land use) coalesced into large farms (regional land use), sedimentation rates increased basin-wide, peaking in the late 19th and early 20th centuries.

Nutrients — Fertilizers and Sewage

Increases of nitrogen and phosphorus in Bay waters from fertilizers closely resemble sedimentation patterns, which is not surprising since these chemicals adhere to silt and clay particles. Measured influxes of phosphorus and total nitrogen preserved in the sediment were compared with historical records of fertilizers purchased in the watershed (Emily Elliott and Dan Bain, unpublished data), and with records since 1900 of nitrogen discharged into the Potomac River (Jaworski et al. 2007). These data show a remarkable similarity between the amount of fertilizers applied to the land and the

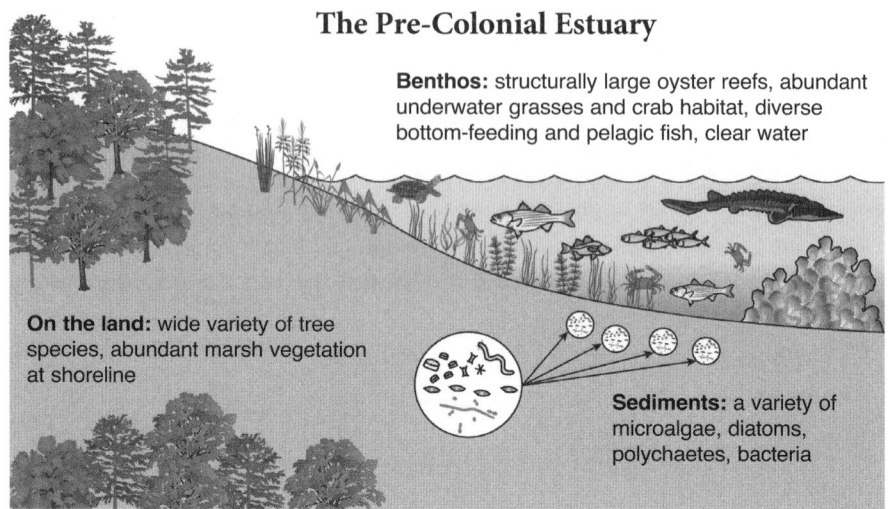

The Pre-Colonial Estuary

Benthos: structurally large oyster reefs, abundant underwater grasses and crab habitat, diverse bottom-feeding and pelagic fish, clear water

On the land: wide variety of tree species, abundant marsh vegetation at shoreline

Sediments: a variety of microalgae, diatoms, polychaetes, bacteria

Figure 13. Sketch of the pre-Colonial estuary, indictating a forested landscape that minimizes runoff of sediments and nutrients, abundant underwater grasses, and a diverse mix of microorganisms and macroorganisms. Illustration by Sandy Rodgers; drawings of plants and organisms from the UMCES Integration and Application Network.

amount of nitrogen and phosphorus reaching estuarine waters. Sherri Cooper showed that carbon, nitrogen, and phosphorus influxes from four cores in a transect across the mid-section of the Bay were similar to patterns of land clearance and sedimentation (Cooper 1993, Cooper and Brush 1991). She also estimated the degree of pyritization of iron, an analysis that allows an approximation of the timing of anoxia in the mesohaline section of the Bay; the analysis showed a rise in anoxia with the input of fertilizers. Using isotopes of nitrogen and sulfur extracted from sediment cores, Emily Elliott identified runoff of manure from cattle as the source of nitrogen in a tributary of the Rhode River, again demonstrating clearly the hydrologic conductivity of land and water (Elliott and Brush 2006). The very early presence of the heavier nitrogen isotope in sediments from the mid Bay, described by Angie Sowers, is also evidence that the earliest Colonial land use was already affecting the chemistry of the main stem of the Bay (Sowers and Brush 2014). The latter is a likely result of the early draining and drying of the land, which would have resulted in decreased denitrification — the release of nitrogen gas into the atmosphere — and subsequent leaching of

The Post-Colonial to Modern-day Estuary

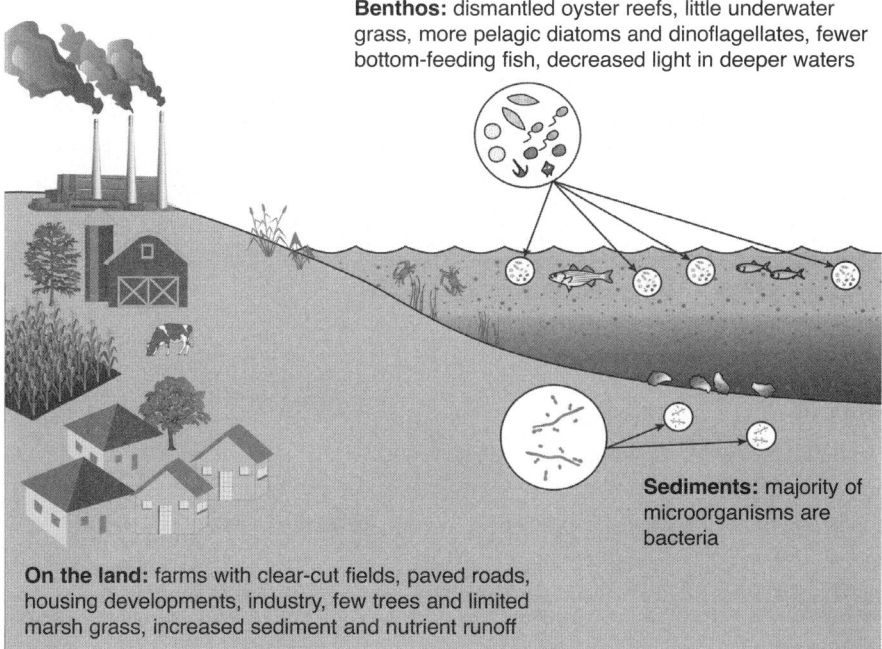

Benthos: dismantled oyster reefs, little underwater grass, more pelagic diatoms and dinoflagellates, fewer bottom-feeding fish, decreased light in deeper waters

Sediments: majority of microorganisms are bacteria

On the land: farms with clear-cut fields, paved roads, housing developments, industry, few trees and limited marsh grass, increased sediment and nutrient runoff

Figure 14. Sketch of the post-Colonial estuary, indicating the effects of forest clearance for farms and urban development. The consequent runoff of soil and nutrients contributes to the loss of underwater grasses and the depletion of dissolved oxygen in bottom waters. Illustration by Sandy Rodgers; drawings of plants and organisms from the UMCES Integration and Application Network.

nitrogen into estuarine waters even before there was increased sedimentation and evidence of eutrophication in diatom populations. Humaira Khan's much earlier studies from Jug Bay showed similar patterns of increased nutrients and metals with land clearance (Khan and Brush 1994).

To demonstrate the effects of nutrients from sewage on aquatic life, we collected sediment cores from Back River and the adjacent Middle River. The Back River Wastewater Treatment Plant (BRWWTP), which serves metropolitan Baltimore, began operating in 1911. The flow of wastewater into Back River was altered twice: in 1940, water was diverted from the BRWWTP via a pipe to the Bethlehem Steel Plant located on Baltimore Harbor (the Patapsco River) to serve as cooling water. In summer 1958,

Bethlehem Steel was not in operation due to a steelworkers' strike and all of the wastewater flowed into Back River. Sedimentary chlorophyll degradation products (SCDP) were measured as a proxy for primary production in the cores collected from Back and Middle Rivers. There was virtually no change in SCDP in Middle River or in the cores collected upstream of the BRWWTP or at the mouth of Back River. However, in cores collected downstream from the BRWWTP, SCDP reflected exactly the wastewater discharge profiles recorded by the BRWWTP: the cores showed a dramatic decrease in SCDP as wastewater was diverted to Bethlehem Steel and a very large peak during the 1958 steelworkers strike. Because the residence time in Back River is approximately 18 days, most of the nutrients are consumed by phytoplankton before reaching the mouth of the tributary. The study demonstrated the rapid response of aquatic organisms to nutrient input (Brush 1984). Sediment cores from Back River also show the complete absence of SAV in the stretch of river below the BRWWTP, again reflecting the effect of extreme eutrophication and anoxia on the living system.

Salinity

Increased runoff from a largely deforested land into the Chesapeake reduced the salinity of estuarine waters. This decreased salinity is shown in sediment cores from the upper tributaries: before colonization, we find seeds of brackish species of SAV; today we find evidence only of seeds of freshwater species (Brush and Hilgartner 2000). Cores from the mouth of the Chester River and the mid Bay show a decrease in sulfur with the onset of colonization, indicating that the increased fresh water from the tributaries was swamping sulfur coming from the tidal waters (Sowers and Brush 2014).

Primary Producers — SAV and Diatoms

In order to reconstruct the history of SAV, an original objective of the Chesapeake coring project, we collected sediment cores in the upper Bay (Furnace Bay), three locations in the middle Bay (Patuxent River), and in the lower Bay (Ware River). Analysis of the cores at all sites showed that the SAV demise in the 1970s was a unique event. While fluctuations occurred in populations prior to the 1970s, the near extinction of all species had not occurred as in recent times. Despite fluctuations in the number of seeds present from time to time, diversity remained fairly constant (Brush et al. 1980, Davis 1985). A more detailed study of SAV based on 34 cores collected in 12 tributaries of the upper and middle Chesapeake showed the effects of local activities on the populations, in particular the BRWWTP. Almost all of these cores showed the SAV die-off in the early 1970s (Brush and Hilgartner 2000).

The historic demise of SAV and the decrease of bottom-dwelling fish correspond to the loss of benthic diatoms and the proliferation of planktonic diatoms recorded in cores from the mid-Bay transect studied by Sherri Cooper as well as in cores collected for studies of SAV. The shift in diatom populations from benthic to planktonic coincides with the increase in sedimentation and chemical elements associated with land clearance and fertilizer use. The response of diatoms to changes on the land was a very large increase in numbers of organisms with a corresponding decrease in diversity (see Figure 15) — nearly the opposite of SAV where numbers decreased drastically, but diversity was not so greatly affected (Cooper 1993, Cooper and Brush 1991). Similar results were shown for the Chester River and the middle lower Bay (Sowers 2003, Sowers and Brush 2014).

Primary and Secondary Consumers — Foraminifera and Worms

Foraminifera, single-celled protozoa with shells, were analyzed in cores from the mid Chesapeake and the mouth of the Chester River: shortly after colonial settlement, a shift in species occurs from those favoring higher salinity or more brackish waters, *Elphidium* sp., to those that prefer fresher water, *Ammobacculites* sp. (Sowers and Brush 2014). *Elphidium* sp. does not occur in the record after colonization.

Analogous findings are evident for detritivores: until colonial settlement, jaws of the polychaete worm *Nereis* sp. were abundant in a core from the mouth of the Chester River; however, when the benthos became anoxic and not habitable for detritivores, *Nereis* populations decreased dramatically. *Nereis* sp. is the preferred food of many of the fish species no longer abundant in the Bay. The abundance of *Nereis* in pre-Colonial sediments suggests that an important transfer of energy in the food web at that time was through organisms that could access nutrients in the sediment.

As the land changed, so eventually did the ecology of the estuary.

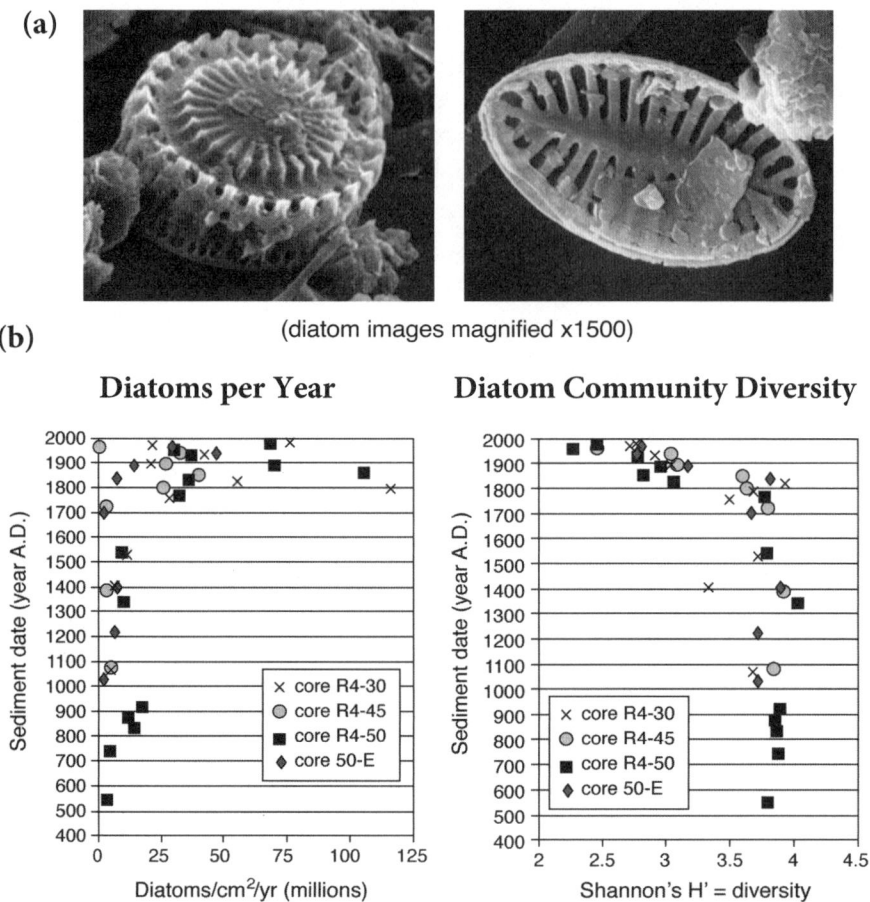

Figure 15. (a) Micrographs of a centric diatom (left) and a pennate diatom (right) and (b) Profiles of diatom abundance and community diversity since 400 AD in a transect across a mesohaline section of the Chesapeake Bay. Redrawn from Cooper and Brush (1993).

Turbidity caused by excess sediment and nutrients originating from the land — and from airborne deposition — reduces the amount of light reaching bottom habitats, thus limiting the ability of SAV and other plants to photosynthesize and produce oxygen in near-bottom waters. An overabundance of nutrients in the upper part of the water column, were light is not limiting, has resulted in a proliferation of algae beyond the grazing capacity of small zooplankton (copepods) and diminished oyster stocks. These ungrazed phytoplankton die and sink into the benthos, where they decompose, a process that consumes oxygen, thus further reducing dissolved oxygen concentrations already compromised by the loss of SAV. Eventually, aerobic organisms and bottom-dwelling fish can no longer live in a benthos starved for oxygen. And the once-rich benthic habitats become "dead zones."

Looking Ahead — The Land and Water

Research aimed at decoding the sediments of Chesapeake Bay has been providing clear evidence of how human activities in the watershed have contributed to changes in the Bay's ecology over the last 400 years. Beyond a greater understanding of this ecosystem, what can these research findings offer natural resource managers who are engaged in diverse restoration activities in the Chesapeake?

John Bonner in *Cells and Societies* characterizes all living societies as having three functions in common: (1) each takes in food so that its individual members may be nourished, (2) each perpetuates itself by reproduction, and (3) each has some coordination, some integration, or communication among its members.

These three activities necessary for a living system — sustenance, continuation, interactions — when applied to the Chesapeake that originated over 10,000 years ago and has been changing ever since, underlie an evolutionary complexity. The Bay's immense productivity over millennia has relied on its ability to adapt to ever-changing environments — from fresh to saline water as sea levels rose; from cold to warm temperatures as glaciers retreated — over time periods that have enabled large organisms with long life spans to adapt genetically or be replaced by organisms with similar life spans.

Adaptation to environmental change may require decades to centuries, especially for larger organisms such as fish and shellfish that are important sources of the seafood economy. But the timeframe for human-induced change is short, compared with climatic and geomorphological changes, and the change favors organisms with very short life cycles such as diatoms and microorganisms. Fish species such as shad, sturgeon, and striped bass declined as they were unable to adjust to rapidly changing estuarine habitats. Since Colonial settlement, the Bay's ecological communities have generally shifted from a benthic system dominated by bottom-dwelling fish and oysters to a pelagic system dominated by microorganisms. While Chesapeake Bay is still a source of fish and shellfish, it is no longer producing seafood as it did when settlers first began harvesting its rich aquatic resources.

A healthy benthos requires water to be "clean" enough so that SAV can thrive and be sustainable, thus providing habitat and food resources for bottom-dwelling species. Therefore, cleaning Bay waters so that bottom-dwelling species can do well has been a prime objective for coordinated restoration efforts since the Chesapeake Bay Program got underway in 1985. All constituents agree that achieving clean water requires programs that will severely limit sediment, fertilizer, and human and animal waste from entering the Bay's riverine waters and reduce significantly vehicular and industrial pollution, including pollutants from the air. Among the practices natural resource agencies are promoting to curtail estuarine pollution are programs designed to use the least amount of fertilizer necessary for crop production; employ cover crops for restoring soil nutrients; upgrade sewage treatment facilities; effectively manage animal waste; and control industrial and automobile pollution. Unfortunately, some of these programs have had limited success because of economic, political, and cultural factors. Among the major constraints are the costs to sustain these programs: who will pay and how? The majority of people who live near the Bay today do not derive their livelihood from fishing; nor does their food consumption depend on the local economy. Their priorities with respect to the health of the Bay ecosystem tend to be more aesthetic and recreational than as a supplier of seafood.

Over time, organisms will follow clean habitats; however, the species may or may not be the ones we have come to know or identify with the Chesapeake, as new genotypes evolve and migrations occur in this open and interconnected system. The recent emergence of SAV in the Potomac River coincident with nutrient reduction is an encouraging sign that species will return to cleaner waters (Ruhl and Rybicki 2010). The re-emergence of large beds of SAV in the Susquehanna Flats in the early 2000s, which had all but disappeared in the early 1970s, also indicates that species will return; however, recovery is complicated because the delivery of nutrients to the estuary is influenced by freshwater discharges that vary with annual precipitation, as well as feedback mechanisms not fully understood (Gurbisz and Kemp 2014).

In the meantime, hatchery and aquacultural practices for rearing oysters so that they can contribute to the economy are being attempted with considerable success. However, in nature, the oyster is dependent on a host of other species in a complicated and interactive living system. The conversion of the living multi-species estuary to a few species, in part through aquaculture or targeted fisheries policies, requires expensive subsidies such as hatcheries and hands-on maintenance of habitats. Such a conversion does little to return the Bay to a benthic system. Estuarine aquaculture production might be likened to the conversion of pre-Colonial forests to agricultural fields. The productivity of the latter is maintained by large "subsidies," i.e., nutrient inputs such as fertilizers and irrigation systems. When farms are abandoned and these subsidies removed, and with no further development, the land over time reverts back to forest.

Our paleoecological research has shown how intricately water and land are related and how changes in land use are reflected in estuarine processes. Our understanding suggests an additional avenue for Bay recovery, which has to do with restoring the land mosaic and the potential it offers for denitrification, which removes nitrogen from the estuary. Restoring the land mosaic includes adaptations of ongoing tree planting activities, which have wide public support.

Studies of Maryland's forest cover indicate a clear relationship between tree species, geology, and soils, and show that the majority of native species are restricted to one substrate, and others to two or a few similar substrates (Brush et al. 1980). For example, chestnut oak grows on coarse schist and gravel, tulip poplar on soils derived from fine schist or gneiss, American holly on sandy soils. Only a few species grow everywhere.

The paleoecological data show that eutrophication and anoxia in the Chesapeake Bay began when more than 60% of the watershed was deforested and small farms separated by woodlots and hedgerows were replaced by large commercial agricultural tracts. We can infer from these data that when the landscape consisted of a mosaic of agricultural fields and wooded areas, there was very little leaching of nitrogen into the soil and ultimately the Bay, even when fertilizers were used. The woodlots and hedgerows consisting of native species acted as buffers throughout the watershed. Preliminary research in a small Baltimore County watershed that analyzed soil samples taken in close proximity to the roots of four different tree species on two different substrates shows that nitrification and mineralization differ not only for different tree species, but for the same species on different substrates (Brush, unpublished data). These data suggest that trees and shrubs could provide effective buffers throughout the watershed. As successful as riparian buffers can be, they are too limited in extent to protect the estuarine system from the entire watershed.

One inference from the data is that small planted woodlots of young tree species on appropriate substrates throughout the watershed, and restoration of some drained areas to their original wet conditions, could simulate the original nutrient cycling, including nutrient retention and denitrification, and help restore clean waters. The woodlots could also contribute to a wood products economy. A key component of such an undertaking would include a significant reduction in fertilizers transported to estuarine waters by runoff and atmospheric deposition.

Over time, as the water becomes cleaner and clearer, we would expect the estuarine ecosystem to gradually reorganize and restore itself to a benthic-pelagic system of producers and consumers with both short- and long-generation times, just as abandoned agricultural fields that are not developed revert over time to a forest of short lived herbaceous plants and trees and shrubs with much longer generation times.

An optimistic scenario!

References

Bain, D.J. 2003. 400 years of land use impacts on landscape structure and riparian sediment dynamics: investigations using chromite mining waste and property mosaics. Ph.D. dissertation, Johns Hopkins University.

Bain, D.J. and G.S. Brush. 2005. Early chromite mining and agricultural clearance: opportunities for the investigation of agricultural sediment dynamics in the Eastern Piedmont (USA). American Journal of Science 305:957-981.

Bain, D.J. and G.S. Brush. 2008. Gradients, property templates, and land use change. The Professional Geographer 60(2):224-237.

Bonner, J.T. 1955. Cells and Societies. Princeton University Press. 234 pp.

Brush, G.S. 1984. Stratigraphic evidence of eutrophication in an estuary. Water Resources Research 20(5):531-547.

Brush, G.S. 1986. Geology and Paleoecology of Chesapeake Bay: a long-term monitoring tool for management. Journal of the Washington Academy of Sciences 76 (3):146-160.

Brush, G.S. 1989. Rates and patterns of estuarine sediment accumulation. Limnology and Oceanography 34:1235-1246.

Brush, G.S. 2009. Historical land use, nitrogen and coastal eutrophication: a paleoecological perspective. Estuaries and Coasts 32: 18-28.

Brush, G.S. and L.M. Brush, Jr. 1972. Transport of pollen in a sediment-laden channel: a laboratory study. American Journal of Science 272:359-381.

Brush, G.S. and L.M. Brush. 1994. Transport and deposition of pollen in an estuary: a signature of the landscape. Pages 33-46 in A. Traverse (ed.) Sedimentation of Organic Particles. Cambridge University Press.

Brush, G.S. and W.B. Hilgartner. 2000. Paleoecology of submerged macrophytes in the upper Chesapeake Bay. Ecological Monographs 70(4):645-667.

Brush, G.S., F.W. Davis, and S. Rumer. 1980. Biostratigraphy of the Chesapeake Bay and its tributaries. Final Report to EPA Chesapeake Bay Studies Program.

Brush, G.S., C. Lenk, and J. Smith. 1980. The natural forests of Maryland: an explanation of the vegetation map of Maryland (with 1:250,000 map). Ecological Monographs 50:77-92.

Colman, S.M., J.P. Halka, C.H. Hobbs III, and R.B. Mixon. 1990. Ancient channels of the Susquehanna River beneath Chesapeake Bay and the Delmarva Peninsula. Geological Society of America Bulletin 102:1268-1279.

Cooper, S.R. 1993. The history of diatom community structure, eutrophication and anoxia in the Chesapeake Bay as documented in the stratigraphic record. Ph.D. dissertation, Johns Hopkins University.

Cooper S.R. and G.S. Brush. 1991. Long term history of Chesapeake Bay anoxia. Science 254:992-996.

Cooper, S.R. and G.S. Brush. 1993. A 2500-year history of anoxia and eutrophication in Chesapeake Bay. Estuaries (16)3B:617-626.

Cronon, L.E. 1986. Chesapeake fisheries and resource stress. Journal of the Washington Academy of Sciences 76(3):188-198.

Davis, F.W. 1985. Historical changes in submerged macrophyte communities of upper Chesapeake Bay. Ecology 66:981-993.

DeFries, R.S. 1980. Effects of land use history on sedimentation in the Potomac Estuary, Maryland. U.S. Geological Survey Water Supply Paper 2234-K. 23 pp.

Elliott, E.M. and G.S. Brush. 2006. Sedimented organic nitrogen isotopes in freshwater wetlands record long-term changes in watershed nitrogen source and land use. Environmental Science and Technology 40(9): 2910-2916.

Gurbisz, C. and W.M. Kemp. 2014. Unexpected resurgence of a large submersed plant bed in Chesapeake Bay: Analysis of time series data. Limnology and Oceanography 59(2):482-494.

Gottschalk, L.R. 1945. Effects of soil erosion on navigation in the upper Chesapeake Bay. Geographical Review 35(2):219-238.

Helms, D. 2000. Soil and Southern history. Agricultural History 74(4):723-758.

Hilgartner, W.B. 1995. Habitat development in a freshwater tidal wetland: a paleoecological study of human and natural influences. Ph.D. dissertation, Johns Hopkins University.

Hilgartner, W.B. and G.S. Brush. 2006. Human impact and late Holocene habitat dynamics in a Chesapeake Bay freshwater tidal wetland delta. Holocene 16(4):479-494.

Jarosewich, M. 1985. Correspondence of historical land use, meteorological events, and sediment accumulation in Furnace Bay, Maryland. M.S. thesis, Johns Hopkins University.

Jaworski, N.A., B. Romano, C. Buchanan, and C.L. Jaworski. 2007. The Potomac River Basin and its Estuary: landscape loadings and water quality trends, 1895-2005. http://www.umces.edu/president/Potomac.

Khan, H. 1993. A paleoecological study of a freshwater tidal marsh. Ph.D. dissertation, Johns Hopkins University.

Khan, H. and G.S. Brush. 1994. Nutrient and metal accumulation in a fresh water tidal marsh. Estuaries 17(2):345-360.

Martin, M. 2011. The evolutionary mechanisms of *Ambrosia artemisiifolia* invasion. Ph.D. dissertation, Johns Hopkins University.

Martin, M.D., E.A. Zimmer, M.T. Olsen, A. D. Foote, M. T. P. Gilbert, and G. S. Brush. 2014. Herbarium specimens reveal a historic shift in phylogeographic structure of common ragweed during native range disturbance. Molecular Ecology 23:1701-1716.

McNown, J., J. Malarka, and H.R. Pramanik. 1951. Particle shape and settling velocity. Internat. Assoc. Hydraulic Research, 4th Mtg. p. 514.

Miller, H. 1986. Transplanting a "Splendid and Delightsome Land": colonists and ecological change in the 17th and 18th century Chesapeake. Journal of Washington Academy of Sciences 76:173-187.

Pasternack, G.B. 1998. Physical dynamics of tidal fresh water delta evolution. Ph.D. dissertation, Johns Hopkins University.

Pasternack, G.B. and G.S. Brush. 1998. Sedimentation cycles in a river-mouth tidal fresh water marsh. Estuaries 21:407-415.

Pasternack, G.B. and L.A. Hinnov. 2003. Hydro meteorological controls on water level in a vegetated Chesapeake Bay tidal fresh water delta. Estuarine, Coastal, and Shelf Science 58(2):373-393.

Pasternack, G.B. and G.S. Brush. 2002. Biogeomorphic controls on sedimentation and substrate on a vegetated tidal freshwater delta in upper Chesapeake Bay. Geomorphology 43:293-311.

Ruhl, H.A. and N.B. Rybicki. 2010. Long-term reductions in anthropogenic nutrients link to improvements in Chesapeake Bay habitat. Proceedings of the National Academy of Science 107(38):16566-16570.

Scatena, F. 1987. Recent stratigraphy of an urban tidal embayment: the Ana-costia River, Maryland. Ph.D. dissertation, Johns Hopkins University.

Shreve, F., M.A. Chrysler, F.H. Blodgett, and F.W. Besley. 1910. The plant life of Maryland. Special Publication Volume III. Maryland Weather Service, The Johns Hopkins Press, Baltimore, Maryland, USA.

Sowers, A. 2003. An interglacial (Sangamon) and late Holocene record of Chesapeake Bay. Ph.D. dissertation, Johns Hopkins University.

Sowers, A. and G.S. Brush. 2014. A paleoecological history of the late pre-Colonial and post-Colonial mesohaline Chesapeake Bay food web. Estuaries and Coasts 37:1506-1515.

Thornton, P. 1992. The effect of rising sea level on salt-marsh development. M.S. thesis, Johns Hopkins University.

Yuan, S. 1993. Postglacial history of vegetation and river channel geomor-phology in a Coastal Plain floodplain. Ph.D. dissertation, Johns Hopkins University.

Afterword — Women Scientists and Paleoecological Research

While *Decoding the Deep Sediments: The Ecological History of Chesapeake Bay* is a scientific story of how paleoecological research has helped inform our knowledge of the Bay's ecosystem changes since Colonial settlement, this Afterword is a personal story about the path that led to doing this research in the first place. The path was by no means a straight one: rather, it followed a convoluted route related to the trajectory of my life from a single to a married woman with three sons. This Afterword covers questions of education and employment opportunities, research funding, the quest for graduate students, child care, and a host of restrictions that circumscribed the role of so many women scientists of my generation.

The author receiving the Mathias Medal. Photograph by Michael W. Fincham.

My introduction to the concept of landscape change over time began with an undergraduate course in paleontology at St. Francis Xavier University in Antigonish, Nova Scotia, where I grew up. Nova Scotia is a narrow peninsula jutting into the Atlantic, a spectacular land of many bogs, lakes, and northern coniferous trees intermingled with maples and aspens along a coast of boulders and sandy beaches. In this course, we visited the Joggins Cliffs on the Bay of Fundy (northwestern side) of the peninsula, where fossil deposits of cycad- and fern-like trees and other plants called *Stigmaria* and *Calamites* going back 300 million years were exposed. The landscape was

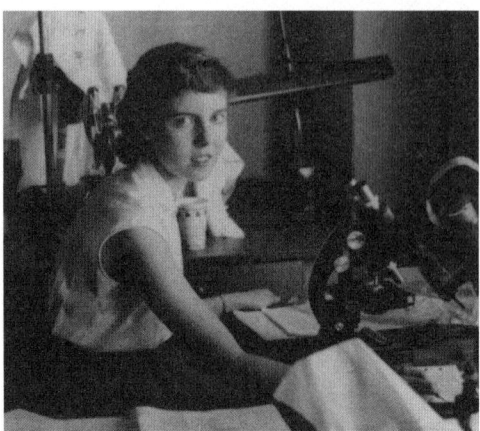

The author as a graduate student at Harvard in 1955. Photograph courtesy of Grace Brush.

like nothing any of us had ever seen. How that ancient landscape changed to what we have today became an underlying question, though often subliminal, in most of my research.

On completing undergraduate studies in 1949 with a major in economics and a minor in geology, the Geological Survey of Canada (GSC) hired me as a technician to work in a small coal research laboratory in Sydney, Nova Scotia. At that time, the coal and steel industry was a large part of Eastern Canada's economy. My job was to prepare small sample blocks of coal, used instead of thin sections, for microscopic studies of the petrography and stratigraphy of coal seams. Under the microscope, I observed odd globular shaped non-crystalline bodies enclosed by discrete walls — these were plant spores. The literature indicated that these fossils could be useful in identifying productive coal seams. With my supervisor's permission, I began studying the palynology (the study of fossil spores and pollen) of the Sydney coal seams. At the end of the year, I presented my preliminary findings at an international conference on Coal Geology in Antigonish. The study was well received by U.S. and European scientists. Based on the potential for palynology as a useful tool for evaluating coal seams, the GSC decided to set up a palynology laboratory.

Since I had begun the work, I was given the opportunity to further my education in coal geology and palynology with the objective of eventually establishing a fossil pollen and spore laboratory to study coalfield stratigraphy in Eastern Canada. I attended the University of Illinois where programs in coal research were being carried out both at the University and the Illinois Geological Survey. Financial support was in the form of a teaching assistantship in the botany department at Illinois and travel funds from the Nova Scotia government. Never having been outside the Canadian Maritime Provinces and only once outside Nova Scotia, the sudden exposure to the flat treeless landscape of at least a part of the Midwestern U.S. seemed unreal, even though as a young child I had devoured books and articles that described the geography of faraway places. The treeless grasslands surrounding Urbana seemed as remote from Nova Scotia as the ancient flora of the Joggins Cliffs.

Palynology is a part of paleobotany and at Illinois, paleobotany was in the botany department; so I was a graduate student in botany.

In addition to learning about spores preserved in coal deposits, I began to study the science of plant evolution. I found this incredibly fascinating and related to my earlier exposure to fossil plants at the Joggins Cliffs.

I completed a M.S. degree, returned to Canada to set up the pollen laboratory but decided after a year to leave the GSC to pursue graduate work in paleobotany. The pollen laboratory was up and running. I began graduate work at Penn State, which had a large program in coal research under the direction of a paleobotanist. I met Lucien Brush who had just graduated from Princeton — he was studying the hydrodynamics and geomorphology of streams. We were married in September of 1953. Once married, my scientific life became complicated with regard to professional opportunities. I was not only a female graduate student unattached to any professional or scientific organization, but I was also married. Shortly after our marriage, Lucien transferred to the geology department at Harvard University as his major advisor was moving there. I was unsure how this move would affect me. The general expectation in science departments was that a married woman would not likely stay the course for a Ph.D., so there was not much incentive for professors (essentially all male at the time) to take them on as students. However, Elso Barghoorn, a paleobotanist in the biology department at Harvard, famous for having discovered the first fossil evidence of life on Earth, was at the conference in Antigonish where I presented my research with the GSC. I contacted him when I knew we would be moving to Cambridge. He accepted me as a student. Those years at Harvard were exceptionally happy ones for both Lucien and me. Barghoorn's laboratory was a center of intellectual thought; furthermore, he made no distinction between male and female students.

My dissertation research was to associate fossil spores in Paleozoic "coal balls" (a type of fossil deposit similar in age to the Joggins fossils) with the plants that produced them, and to trace the history of the plants from the spores preserved in these ancient rocks. Scientifically, and intellectually, everything had come together with respect to my research interests and I completed a Ph.D. in biology. However, when we left Harvard in 1956, I did not know what I would be doing next, because we had decided that the practical thing for us would be to go wherever Lucien got the best available position and I would figure out how to continue my work — not unlike the situation today, but now for men as well as women.

Since our fields were quite different, the chances of finding places with both hydrodynamics and paleobotany programs were not high. Barghoorn advised me as I was leaving Harvard to do my own research and not succumb to working as an assistant for someone else. Women who pursued graduate studies in those days often did not complete dissertations and even when they did, many ended up as technicians for their husbands who they met in graduate school. It was easier for Barghoorn to give me this advice than for me to carry it out. But he was right! And insofar and whenever I was able to get support, and even without support, I have done my own research.

Lucien's profession took him first to the U.S. Geological Survey in Washington, D.C., and Denver, Colorado, then to the University of Iowa, later to Princeton, and finally to Johns Hopkins University in 1969. Each of those institutions was extremely generous in making laboratory space available to me for carrying out paleobotanical and palynological studies that required laboratories equipped with a fume hood and a separate space for microscopy. In the late 1950s and the 1960s, I received National Science Foundation grants through the University of Iowa and Princeton University to study the behavior of pollen transport in water — basic studies for interpreting pollen profiles. I was questioned at times by department chairmen about asking for salary money on NSF proposals, one comment being, "You are married so why are you asking for salary for yourself?" I tried to justify my request by pointing out that I needed money for childcare while working in the field or the laboratory. For a large part of my professional life I was supported by a much-appreciated provision of university laboratory and office space and periodic grants and contracts from state, federal, and private agencies. In 1990, at the age of 59, I became a tenured full professor at Johns Hopkins, at which time I began to receive a salary for the courses I taught and the graduate students I advised.

Juggling childcare and working was always a challenge because there were essentially no day-care facilities during those early years. Since my hours were fairly flexible, I could avail myself of student services like the

Tiger Tot Tenders at Princeton. At other places where we lived, there were various makeshift arrangements, some of which were great and others not so good. Finding child care was never easy and always an overriding worry. Sad to say, it is still a great concern — both in availability and cost. For example, an excellent day care facility recently made available on the Johns Hopkins Homewood campus costs annually in excess of $20,000 per child, and although this cost is standard for such facilities, it is unaffordable for many parents.

None of my research would have been accomplished without Lucien's support, which included above all interest in and encouragement of my studies. It is hard today to realize the importance and uniqueness of the support from my husband during the 1950s, 1960s, and later. I could not have continued my research otherwise.

I am also very fortunate in the graduate students who are attracted to the kind of research I do. They have gone on to successful careers in the environmental and ecological arena. And although most are not doing paleoecological research, I like to think that paleoecology has had some influence on their approach to environmental issues.

Appendix

A. Literature Resulting from Projects That Form the Basis of This Book*

Watershed Studies

Bain, D.J. 2003. 400 years of land use impacts on landscape structure and riparian sediment dynamics: investigations using chromite mining waste and property mosaics" Ph.D. dissertation, Johns Hopkins University.

Bain, D.J. and **G.S. Brush**. 2004. Placing the pieces: reconstructing the original property mosaic in a warrant and patent watershed. Landscape Ecology 19:843-856.

Bain, D.J. and **G.S. Brush**. 2005. Early chromite mining and agricultural clearance: opportunities for the investigation of agricultural sediment dynamics in the Eastern Piedmont (USA). American Journal of Science 305:957-981.

Bain, D.J. and **G.S. Brush**. 2008. Gradients, property templates and land use change. The Professional Geographer 60(2):224-237.

Brush, G.S. 1982. Environmental analysis of forest patterns. American Scientist 70: 18-25.

Brush, G.S. 2001. Forests before and after the colonial encounter. Pages 40-59 in: P. Curtin, G.S. Brush and G. Fisher (eds.) "Discovering the Chesapeake." Johns Hopkins University Press, Baltimore, Md.

Brush, G.S., **C. Lenk** and **J. Smith**. 1976. A vegetation map of Maryland: the existing natural forests, scale 1:250,000. Williams and Heintz, Washington, D.C.

Brush, G.S., **C. Lenk**, and **J. Smith**. 1980. The natural forests of Maryland: an explanation of the vegetation map of Maryland (with 1:250,000 map). Ecological Monographs 50:77-92.

Cadenasso, M.L., S.T.A.Pickett, L.E. Band, **G.S. Brush**, M.F. Galvin, P.M. Groffman, J.M. Grove, G. Hagar, V. Marshall, B. McGrath, J. O'Neil-Dunne, B. Stack, and A. Troy. 2008. Exchanges across land-water-scape boundaries in urban systems: strategies for reducing nitrate pollution. Annals of New York Academy of Sciences 1134:213-232.

Groffman, P.M., **D.J. Bain**, L.E. Band, K.T. Belt, **G.S. Brush**, J.M. Grove, R.V. Pouyat, I.C. Yesilonis, and W.C. Zipperer. 2003. Down by the riverside: urban riparian ecology. Frontiers in Ecology and the Environment 1(6):315-321.

Groffman, P.M., R.V. Pouyat, M.L. Cadenasso, W.C. Zipperer, K. Szlavecz, I.D. Yesilonis, L.E. Band, and **G.S. Brush**. 2006. Land use context and natural soil controls on plant community composition and soil nitrogen and carbon dynamics in urban and rural soils. Forest Ecology and Management 236:177-192.

Martin, M.D., E.A. Zimmer, M.T. Olsen, A.D. Foote, M.T.P. Gilbert, and **G.S. Brush**. 2014. Herbarium specimens reveal a historic shift in phylogeographic structure of common ragweed during native range disturbance. Molecular Ecology doi:10.1111/mec.12675.

Pickett, St.A., **G.S. Brush**, A.J. Felson, B.P. McGrath, J.M. Grove, C.H. Nilon, K. Szlavecz, C. Swan, and P.S. Warren. 2012. The Baltimore Ecosystem Study: understanding and working with urban biodiversity. Citigreen 4:68-77.

---

* *Names in **bold** are Johns Hopkins University researchers involved in these studies.*

Estuarine Studies

Brush, G.S. 1984. Stratigraphic evidence of eutrophication in an estuary. Water Resources Research 20(5):531-541.7

Brush, G.S. 1984. Patterns of recent sediment accumulation in Chesapeake Bay (Virginia-Maryland, USA) tributaries. In: J.A. Robbins (Guest Editor), Geochronology of Recent Deposits. Chemical Geology 44:227-242.

Brush, G.S. 1986. Geology and paleoecology of Chesapeake Bay: a long-term monitoring tool for management. Journal of the Washington Academy of Sciences (Special Volume in the Historical Perspective of the Estuary in Management) 76(3):146-160.

Brush, G.S. 1989. Rates and patterns of estuarine sediment accumulation. Limnology and Oceanography 34:1235-1246.

Brush, G.S. 1989. Retrieving environmental data. A review of Chesapeake Environmental Data Directory. Ecology 70:527.

Brush, G.S. 1991. The stratigraphic history of water quality in the Chesapeake Bay. Geotimes. 36(12):21-23.

Brush, G.S. 1994. Human impact on estuarine ecosystems: an historical perspective. Pages 397-416 in: N. Roberts (ed.), Global Environmental Change: Geographical Perspectives: Blackwell Publishing Company.

Brush, G.S. 1997. History and impact of humans on Chesapeake Bay. Pages 125-145 in: R.D. Simpson and N.L. Christensen, Jr. (eds.) Ecosystem Function and Human Activities: Reconciling Economics and Ecology. Chapman & Hall, New York, NY.

Brush, G.S. 1999. Ecological consequences of soil erosion and sedimentation. International Journal of Sediment Research, Special Issue on Sediment Transport and Disasters 14(2):355-361

Brush, G.S. 2001. Natural and anthropogenic changes in Chesapeake Bay during the last 1000 years. International Journal of Human and Ecological Risk Assessment 7(5):1283-1296.

Brush, G.S. 2007. Contributing author: Roundtree, H.C., W.E. Clark, and K. Mountford. John Smith's Chesapeake Voyages 1608-1609. University of Virginia Press.

Brush, G.S. 2009. Historical land use, nitrogen and coastal eutrophication: a paleoecological perspective. Estuaries and Coasts 32:18-28 DOI 10.1007/s12237-008-9106-z (selected by Springer for "open access").

Brush, G.S. and **F.W. Davis**. 1984. Stratigraphic evidence of human disturbance in an estuary. Quaternary Research 22:91-108.

Brush, G.S. and **R.S. DeFries**. 1981. Spatial patterns of pollen in surface sediments of the Potomac estuary. Limnology and Oceanography 26(2):295-309.

Brush, G.S. and **W.B. Hilgartner**. 2000. Paleoecology of submerged macrophytes in the upper Chesapeake Bay. Ecological Monographs 70(4):645-667.

Brush, G.S., E.A. Martin, **R.S. DeFries**, and C.A. Rice. 1982. Comparisons of 210Pb and Pollen methods for determining rates of estuarine sediment accumulation. Quaternary Research 18(2):196-217.

Canuel, E.A , **G.S. Brush**, T.M. Cronin, R. Lockwood, and A.R. Zimmerman. 2011. Paleoecology in Chesapeake Bay: a model system for understanding interactions between climate, anthropogenic activities and the environment. In John Gibson (ed.) Applications of Paleoenvironmental Techniques in Estuarine Studies. Springer Developments in Paleoenvironmental Research Series (in press).

Cooper, S.R. 1995. Chesapeake Bay watershed historical land use: impact on water quality and diatom communities. Ecological Monographs 5:703-723.

Cooper, S.R. and **G.S. Brush**. 1991. Long-term history of Chesapeake Bay anoxia. Science 254:992-996.

Cooper, S.R. and **G.S. Brush**. 1993. A 2,500-year history of anoxia and eutrophication in Chesapeake Bay. Estuaries (16)3B:617-626.

Davis, F.W. 1985. Historical changes in submerged macrophyte communities of upper Chesapeake Bay. Ecology 66:981-993.

DeFries R. 1986. Effects of land use history on sedimentation in the Potomac estuary. USGS Water Supply Paper 2234-K. 23 pp.

Elliott, E.M. and **G.S. Brush**. 2006. Organic nitrogen isotopes record long-term changes in watershed nitrogen sources and land use. Environmental Science and Technology 40(9):2910-2916.

Fletcher, C.H., J.E. Van Pelt, **G.S. Brush**, and J. Sherman. 1993. Tidal wetland record of Holocene sea-level movements and climate history. Palaeogeography, Palaeoclimatology, Palaeoecology 102:177-213.

Gorham, E., **G.S. Brush**, L.J. Graumlich, M.L. Rosenzweig, and A.H. Johnson. 2001. The value of paleoecology as an aid to ecosystem monitoring. Environmental Reviews. 9(2):99-126.

Hilgartner, W.B. and **G.S. Brush**. 2006. Human impact and late Holocene habitat dynamics in a Chesapeake Bay freshwater tidal wetland delta. Holocene 16(4):479-494.

Kemp, W.M. and 18 authors including **G.S. Brush**. 2006. Eutrophication of Chesapeake Bay: historical trends and ecological interactions. Marine Ecology Progress Series 303:1-29.

Khan, H. and **G.S. Brush**. 1994. Nutrient and metal accumulation in a freshwater tidal marsh. Estuaries (17)2:345-360 .

Pasternack, G.B. and **G.S. Brush**. 1998. Sedimentation cycles in a river-mouth tidal fresh water marsh. Estuaries 21:407-415.

Pasternack, G.B. and **G.S. Brush**. 2002. Spatial patterns of biogeomorphic controls on sedimentation and substrate on a vegetated tidal freshwater delta in upper Chesapeake Bay. Geomorphology 43:293-311.

Pasternack, G.B., **G.S. Brush**, and **W.B. Hilgartner**. 2001. Impact of historic land-use change on sediment delivery to an estuarine delta. Earth Surface Processes and Landforms 26:409-427.

Pasternack, G.B., **W.B. Hilgartner**, and **G.S. Brush**. 2000. Biogeomorphology of an upper Chesapeake Bay river-mouth tidal freshwater marsh. Wetlands 20(3):520-537.

Sowers, A.A. and **G.S, Brush**. 2014. A paleoecological history of the late pre-Colonial and Postcolonial mesohaline Chesapeake Bay food web. Estuaries and Coasts 37:1506-1515.

Pollen Transport Studies

Brush, G.S. and L.M. Brush, Jr. 1972. Transport of pollen in a sediment-laden channel: a laboratory study. American Journal of Science 272:359-381.

Brush, G.S. and L.M. Brush. 1994. Transport and deposition of pollen in an estuary: a signature of the landscape. Pages 33-46 in A. Traverse (ed.). Sedimentation of Organic Particles. Cambridge University Press.

Martin, M.D., M. Chamecki, and **G.S. Brush**. 2010. Anthesis synchronization and floral morphology determine diurnal patterns of ragweed pollen dispersal. Journal of Agriculture and Forest Meteorology 150:1307-1317.

Martin, M.D., E.A. Zimmer, M.T, Olsen, A.D. Foote, M.T.P. Gilbert, and **G.S. Brush**. 2014. Herbarium specimens reveal a historical shift in phytogeographic structure of common ragweed during native range disturbance. Molecular Ecology 23:1701-1716.

Martin, M.D., M. Chamecki, **G.S. Brush**, C. Meneveau, and M.B. Parlange. 2009. Pollen clumping and wind dispersal in an invasive angiosperm. American Journal of Botany 96(9):1-9. DOI: 10-3732/ajb.0800407.

van Hout, R., M. Chamecki, **G.S. Brush**, J. Katz, and M. Parlange. 2008. The influence of local meteorological conditions on the circadian rhythm of corn (*Zea mays* L.) pollen emission. Agricultural and Forest Meteorology 148:1078-1092.

B. Books*

Brush, G.S. 2007. Contributing author: In Roundtree, H.C., W.E. Clark, and K. Mountford. John Smith's Chesapeake Voyages 1608-1609. University of Virginia Press.

Curtin, P.D., **G.S. Brush**, and G.W. Fisher (eds.). 2001. Discovering the Chesapeake: The History of an Ecosystem. Johns Hopkins University Press, Baltimore, Maryland.

C. Reports*

Brush, G.S. 1984. Recent Diatom and Trace Metal Distributions in the Patuxent and Nanticoke Rivers. Maryland Power Plant Siting Program Final Report.

Brush, G.S., **F.W. Davis**, and **S. Rumer**. 1980. Biostratigraphy of the Chesapeake Bay and Its Tributaries. Final Report to EPA Chesapeake Bay Studies Program.

Brush, G.S., **F.W. Davis**, and **C.A. Stenger**. 1982. Sediment Accumulation and the History of Submerged Aquatic Vegetation and Eutrophication in the Patuxent and Ware Rivers: A Stratigraphic Study. Final Report. US EPA Chesapeake Bay Program.

Brush, G.S., J.M. Hill, and M.A. Unger. 1998. Pollution History of the Chesapeake Bay. NOAA Technical Memorandum NOS ORCA 121, Final Report.

NRC Committee to Evaluate Indicators for Monitoring Aquatic and Terrestrial Environments. 2000. Ecological Indicators for the Nation. National Academy Press.

D. Supervised Dissertations and Theses

2011. Michael D. Martin, Ph.D. The Evolutionary Mechanisms of *Ambrosia artemisiifolia* Invasion.

2005. Joseph Smith, M.S., Influence of Channel Morphology on Vegetation in the Gwynns Falls Watershed.

2004. Mason Throneburg, M.S. Geomorphic-hydrologic Vegetation Relationships in the Gwynns Falls Watershed.

2003. Daniel Bain, Ph.D. 400 Years of Land Use Impacts on Landscape Structure and Riparian Sediment Dynamics: Investigations Using Chromite Mining Waste and Property Mosaics.

2003. Emily Elliott, Ph.D. Organic Nitrogen Isotope Stratigraphy, Palynology and Sediment History of Freshwater Wetlands in the Chesapeake Bay Basin: Comparison with Land Use History.

2003. Angela Arnold Sowers, Ph.D. An Interglacial (Sangamon) and Late Holocene Record of Chesapeake Bay.

1998. Gregory Pasternack, Ph.D. Physical Dynamics of Tidal Freshwater Delta Evolution.

1995. William Hilgartner, Ph.D. Habitat Development in a Freshwater Tidal Wetland: A Paleoecological Study of Human and Natural Influences.

1995. Shaomin Yuan, Ph.D. Postglacial History of Vegetation and River Channel Geomorphology in a Coastal Plain Floodplain.

1993. Sherri Cooper, Ph.D. The History of Diatom Community Structure, Eutrophication and Anoxia in the Chesapeake Bay as Documented in the Stratigraphic Record.

1993. Humaira Khan, Ph.D. A Paleoecological Study of a Freshwater Tidal Marsh.

1992. Peter Thornton, M.S. The Effect of Rising Sea Level on Salt-marsh Development.

1991. Andrea Buckley, M.S. Analyses of a Sediment Core from a Cypress Swamp in Delaware with a Description of the Habitat Succession.

1987. Fred Scatena, Ph.D. Recent Stratigraphy of an Urban Tidal Embayment: The Anacostia River, Maryland.

1985. Martha Jarosewich, M.S. Correspondence of Historical Land Use, Meteorological Events, and Sediment Accumulation in Furnace Bay, Maryland.

1984. Candice Wilderman, Ph.D. The Floristic Composition and Distributional Patterns of Diatom Assemblages in the Severn River, Maryland.

1983. Georgia Marino, M.S. Environmental Reconstruction of the Upper Potomac Estuary.

1982. Frank W. Davis, Ph.D. The History of Submerged Aquatic Vegetation at the Head of Chesapeake Bay: A Stratigraphic Study.

1982. Cynthia S. Stenger, M.S. A Palynological Study of Sediments from the Chesapeake Bay Area.

1981. Donald G. Parker, M.S. Seed Dispersal of Black Willow (*Salix nigra*) Marsh.

1980. Ruth S. DeFries, Ph.D. Sedimentation Patterns of the Potomac Estuary Since European Settlement: A Palynological Approach

1979. L. Reed Huppman, M.S. Distributional Patterns of Some Herbaceous Plants in Piedmont Forests, Maryland.

1972. Stephen L. Arnold, M.S. Modern Pollen in the Waters of the Northern Chesapeake Bay.

Note: Names of undergraduates who contributed to these studies as advisees and research assistants range in the hundreds and are not recorded here.

Acknowledgments

I wish to thank especially Fredrika Moser for proposing the idea of writing this monograph in the first place, Merrill Leffler and Fredrika who carefully and patiently edited several versions of the manuscript and offered much-needed suggestions particularly with regard to organization, and Sandy Rodgers for her editorial work and design. Sue McLaughlin, a long-time friend, read an early version and offered advice from the viewpoint of a non-scientist. Emily Elliott, Dan Bain, Frank Davis, and Bill Hilgartner made corrections to an earlier version. A discussion with Walter Boynton about some of the ideas presented with respect to restoration was most helpful. Thanks to all.

I also acknowledge and thank the many scientists, managers, and students who worked on the projects discussed here. The research was supported by numerous federal and state agencies, private foundations, and consulting companies, incuding the National Science Foundation, the National Oceanic and Atmospheric Administration, the U.S. Environmental Protection Agency, the U.S. Geological Survey, the U.S. Forest Service, Maryland Sea Grant, Maryland Geological Survey, Maryland Department of Natural Resources, the Mellon Foundation, and Louis Bergere Associates.

About This Monograph Series

This monograph is part of a series entitled *Chesapeake Perspectives*, produced by the University of Maryland Sea Grant College to encourage researchers, scholars, and other thinkers to share their insights into the unique culture and ecology of the Chesapeake Bay. Its audience includes environmental scientists and scholars, from marine biologists to cultural anthropologists, and a broad interested public that encompasses resource managers, watershed organizations, and citizen advocates. For more about books in the series and related topics, visit the web at www.mdsg.umd.edu.